Construction Law:
Contracts, Risks and Regulations

Second Edition

Construction Law:
Contracts, Risks and Regulations
Second Edition

Gregory F. Starzyk

Civilisation Chronicles
2015

Copyright © 2015 by Gregory F. Starzyk

All rights reserved. This book or any portion thereof may not be reproduced or used in any manner whatsoever without the express written permission of the copyright holder except for the use of brief quotations in a book review or scholarly journal.

First Printing: 2014
Second Edition: 2015

ISBN 978-0-9907394-1-8

Graphic illustrations by Cynthia L. Lambert

Published under license by Civilisation Chronicles
San Luis Obispo, CA

Disclaimer:
This publication is designed to provide accurate and authoritative information in regard to the subject matter covered. It is sold with the understanding that neither author nor publisher is engaged in rendering legal, accounting, or other professional services. If legal advice or other expert assistance is required, the services of a competent professional person should be sought.

Ordering information:
Special discounts are available on quantity purchases by corporations, associations, educators and others. For details, contact the publisher at civilisationchronicles@gmail.com

Dedicated to Christina, Kathryn and Kenneth

Contents

Foreword xvii
Preface xx
Part I – Construction Contracts 1
Chapter 1   Axioms of Construction Contract Law 3
  1.1   **Introduction** 3
  1.2   **Meaning and Reason for Contract** 4
  1.3   **Sources of Law** 5
          Statutory Law 5
          Common Law 6
          State Law 6
  1.4   **Consideration** 6
          Donative Promises 7
          Bad Bargains 9
    1.4.1   Mutual Assent 9
          Meeting of the Minds 10
          Implied-In-Fact Contracts 11
          Bilateral and Unilateral Agreements 11
    1.4.2   Verbal Contracts 12
          The Statute of Frauds 13
  1.5   **Remedies** 13
          Damages 13
          Restitution 14
          Unjust Enrichment 15
          Quantum Meruit 16
  1.6   **Agreements to Agree** 17
          Letter of Intent 17
          Memorandum of Understanding 18
          Inchoate Contracts 18
          Enforceability of Agreements to Agree 18
  1.7   **Implied Warranty** 19
          Workmanlike Performance 19
          Warranty of Fitness 20

	1.8	Good Faith and Fair Dealing	20
		Uncovering Design Problems	21
		Duty of Cooperation	21
		Reporting Errors or Omissions	22

Chapter 2   Written Contract Form & Substance 23

	2.1	Introduction	23
	2.2	The Fully Integrated Agreement	24
		The Contract Documents	25
		The Work of the Contract	28
	2.3	Roles and Responsibilities	29
		The Owner	30
		Construction by Owner or Separate Contractors	31
		The Contractor	31
		Performance	32
		Shop Drawings, Product Data & Submittals	34
		Protection of Persons and Property	35
		The Designer	37
		Administration of the Contract	37
		Subcontractors	38
		Assignment	39

Chapter 3   Payments & Performance 41

	3.1	Introduction	41
	3.2	Payment Processes	42
		Schedule of Values	42
		Application for Payment	42
		Certificate for Payment	43
	3.3	Pricing Methods	44
		Fixed-Price	44
		Cost-Plus	45
		Fiduciary Duties and Cost-Plus	46
		Federal Acquisition Regulation & Cost-Plus	47
		Unit-Price	48
	3.4	Performance	49
		Substantial Completion	50
		Final Completion	51
		Substantial Completion and Liquidated Damages	53

Chapter 4   Concealed or Unknown Conditions 55

- 4.1   Introduction 55
- 4.2   **Historical Background** 55
  - Paradine v. Jane 56
  - Liability 56
  - The *Spearin* Doctrine 57
- 4.3   **Modern Rules** 58
  - Type I Differing Conditions 59
  - Type II Differing Conditions 60
  - Allocation of Liability 60
  - Disclaimers 61
  - Burial Grounds, Archaeological Sites & Wetlands ... 62

Chapter 5   Changes 63

- 5.1   Introduction 63
- 5.2   **The Changes Clause** 63
  - Cardinal Change 64
  - Contractor-Subcontractor Agreements 65
  - Practical Advice 65
- 5.3   **Contractual Processes** 65
  - Change Orders 66
  - Construction Change Directives 66
  - Minor Changes in the Work 67
- 5.4   **Preexisting Duty Rule** 67
  - The Uniform Commercial Code 68

Chapter 6   Time 69

- 6.1   Introduction 69
- 6.2   **Performance** 69
  - Security Interests 70
  - Late Performance 71
  - Liquidated Damages 72
  - The Anomalies of Time 73
- 6.3   **Liabilities** 74
  - Delay 74
    - Excusable Delay 74
    - Force Majeure 75
    - Compensable Delay 76
    - Concurrent Delay 76
    - Recovery for Compensable Delay 78

    Acceleration 79
    No Damages for Delay 79

Chapter 7 Interpreting the Contract Documents 81

7.1 Introduction 81
7.2 Interpretive Evidence 81
    Lists 83
    Designer's Authority to Interpret the Design 84
7.3 Harmony 84
7.4 Conflicting Information 85
    Order of Precedence 85
    Ambiguity 87
    Contra Proferentum 87
7.5 Misleading Information 88
    Prescriptive Specifications 88
    Performance Specifications 90

Chapter 8 Suspension and Termination 91

8.1 Introduction 91
8.2 Conditions 92
    Legal Conditions 92
    Failed Conditions 93
8.3 Breached Promises 94
    Material Breach 94
8.4 Suspension and Termination for Cause 97
    Suspension 97
     Suspension by the Owner for Cause 97
     Suspension by the Contractor 97
    Termination 98
     Termination by the Contractor 98
     Termination by the Owner for Cause 100
8.5 Suspension or Termination for Convenience 102
    Suspension by the Owner for Convenience 102
    Termination by the Owner for Convenience 102
8.6 Delay and its Implication on Material Breach 103
8.7 Consequential Damages 104

Chapter 9 Assurance 105
9.1 Introduction 105
9.2 Warranty 106
Warranty and Material Breach 107
Construction Defects 108
9.3 Inspection 108
Inspection of Goods 109
Constructive Change 110
9.4 Correction of Work 111
Statutes of Limitations 111
One-Year Correction Period 112

Chapter 10 Claims and Dispute Resolution 115
10.1 Introduction 115
10.2 Claims 115
The Initial Decision 116
10.3 Dispute Resolution 118
Dispute Review Boards 119
Project Neutral 120
Mini-Trial 120
Summary Jury Trial 120
Mediation 121
Arbitration 122
Litigation 123
Arbitration vs. Litigation 125
Federal Contract Disputes 127

Part II – Construction Risks 129

Chapter 11 Bidding on Public Projects 131
11.1 Introduction 131
11.2 Bidding processes 133
Responsive Bidder 133
Responsible Bidder 135
Competitive Sealed Bidding 137
Best Value Procurement 138

	Project Delivery Strategies	140
	Design-Bid-Build	140
	Design-Build	141
	Construction Management	142
	Public Private Partnership	145
	Job Order Contracting	146
	e-Bidding	147
11.3	**Bid Irregularities**	**148**
	Unilateral Mistakes	148
	Mistakes Noticed Before Award	149
	Mutual Mistake	150
	Mistakes Noticed After Award	152
	Bid Mistakes by Subcontractors	152
	Bid Protests to Public Agencies	153
Chapter 12	**Subcontractor Protections & Vulnerabilities**	**157**
12.1	**Introduction**	**157**
	Assignment vs. Subcontracting	157
	Managing Risk	158
	Communication: Contractor/Subcontractor Relationship	159
12.2	**Subcontracting Issues**	**160**
	Flow-thru Clauses and Conflicting Documents	160
	Owner Selection and Approval	161
	Bargaining Disparity	162
	Delayed Payment and Non-Payment	164
	Pay-when-Paid Clauses	165
	Retention	165
	Changes Clause in Subcontractor Agreements	166
12.3	**Mechanics' Liens**	**166**
	Stop Notices	168
	Statutory Lien Processes	169
	Lien Waivers	170
	Special Payment Considerations for Subs	170

Chapter 13   Ethical Considerations for Constructors 173
    13.1   Introduction 173
    13.2   A Code of Ethics 174
                Bidding for Work 174
                      Falsities 174
                      Kickbacks 175
                      Collusion 176
                Other Code of Ethics Topics 177
                      Subcontractor and Supplier Relationships 177
                      Capabilities and Competencies 177
                      Conflict of interest 178
                      Safety, Employment and Compliance 179
                      Professional Stature 180
                      Enforcement 181
    13.3   Bid Shopping and Bid Peddling 181
                Irrevocability of Subcontractor Bids 183
                Bid Shopping an Irrevocable Subcontractor Bid ... 184
                Reverse Bid Auctions 185

Chapter 14   Construction Insurance 187
    14.1   Introduction 187
    14.2   Axioms of Liability 188
                Vicarious Liability 188
                Negligence 189
                      Duty 190
                      Breach 190
                      Cause 190
                      Harm 191
                Strict Liability 192
                Relief from Liability 193
                      Contributory Negligence 193
                      Comparative Negligence 193
                      Assumption of Risk 194
                      Sovereign Immunity 195
                Indemnification 195
                Subrogation 197
                The Fiduciary 197

14.3 Insurance Products 198
       Commercial General Liability Insurance 199
       Builders Risk Insurance 201
       Equipment Floater Insurance 203
       Automobile Insurance.. 203
       Umbrella Excess Liability Insurance 204
       Workers' Compensation Insurance 204

Chapter 15   Surety Bonds 207

15.1 Introduction 207
15.2 The Miller Act 208
       Performance Bonds... 208
       Payment Bonds 210
15.3 Other Surety Bonds 211
       Bid Bond 211
       License Bond... 212
       Lien Release Bond 213
15.4 Surety Transactions 214
       Surety Investigations.. 214
       Indemnification Agreements..................................... 216
       Bond Premiums 216

Part III – Construction Regulations.. 217

Chapter 16   Licensing and Entitlement 219

16.1 Introduction 219
16.2 Contractor Licensing 219
       Contractor Licensing in California 221
       Enforcement.. 224
16.3 Statutory Proscriptions to Contract 225
       Unlicensed California Contractors 227
16.4 Entitlement 228
       Prevailing Wages... 228
       Set-Aside 228
         Section 8(a) Program... 230
         Small Disadvantaged Business Program 230
         Other Notable Set-Asides 231
       Affirmative Action 231
       Domestic Materials Purchasing 232

Chapter 17 Environmental Law 233
   17.1 Introduction 233
   17.2 Asbestos 234
         Asbestos Risk 234
         California Regulatory Scheme 235
         Renovation Projects 236
         Demolition 237
   17.3 Radon 237
         California's Indoor Radon Program 237
         Radon-Resistant Building Techniques 238
   17.4 Lead-Based Paint 238
         Federal RRP Program 239
         Pre-Renovation Education Requirements 240
         Training, Certification & Work Practice 241
         California's Lead-Based Paint Rules 241
   17.5 Polychlorinated Biphenyls 243
   17.6 Storm Water Pollution Prevention Plans 245

Chapter 18 Labor Law 247
   18.1 Introduction 247
   18.2 The U.S. Labor Movement 247
         Recent Trends 249
         Legal Responses 250
   18.3 The National Labor Relations Act 250
         Historical Background 250
         Overview 253
         The National Labor Relations Board 254
            Organization 254
            Jurisdiction 255
   18.4 The Unionization Process 256
         Union Certification 257
         Unfair Labor Practices 259
   18.5 Labor Disputes 261
    18.5.1 Strikes and Lockouts 261
    18.5.2 Picketing 262
         Secondary Boycotts 263
            Common Situs Picketing 264
         Jurisdictional Picketing 265
         Hot Cargo Clauses 265

18.5.3 Remedies 265
National Emergencies 266
18.6 Labor Agreements 266

Chapter 19 Employment Law 269

19.1 Introduction 269
Background 269
Employment-at-Will 270
19.2 Employment Discrimination 272
Disparate Treatment 273
Direct Evidence 274
Facially Discriminatory Policies or Practices 274
Reverse Discrimination 274
Pretext 275
Mixed Motive 275
Pattern or Practice 275
Harassment 276
Adverse Impact 276
Failure to Reasonably Accommodate 277
Retaliation 277
19.2.1 Equal Pay Act 278
19.3 Fair Labor Standards 279
The White Collar Exemption 280
Compensation Categories 282
19.4 Occupational Safety and Health 283
The Occupational Safety and Health Act 283
OSHA Standards 284
OSHA Inspections 284
Recordkeeping 285
Workplace Injuries 286
19.5 Enforcement 286
Arbitration Agreements 288

References 289
References to Legal Materials 295
Index 303

FOREWORD

Having been involved in the construction industry for many years, first as a field installer, then an estimator, a project manager and now as a chief executive, I have benefited both from writings on construction law topics and from personal counsel. Written materials always help to define legal terms and to clarify legal issues. But when written materials fall short, as they sometimes do, personal counsel is necessary. Personal counsel, however, is a last resort since it is expensive and it may be too time consuming to reach a particular objective.

Both the seasoned veteran and neophyte construction person is well served with an instructional resource that touches on the many legal and contractual challenges and outlines preventative measures to avoid future legal effort. Greg`s *Construction Law* textbook is an excellent instructional resource on law and contracts. One does not have to look further as the book provides sufficient clarity. It also fosters thoughtful evaluation should deeper specific legal interpretation or definition be warranted. An example is the explanation of the contractual term "time is of the essence," and how the contract drafter can use that to its advantage. Another is the articulation of the *Spearin* doctrine and how that can place design responsibility on the owner and not on the contractor should there be performance issues.

Greg's book is a must for the entry level person in all disciplines of construction, whether in the general or specialty contracting arena; for the construction and engineering student; and for the non-student entering the industry.

I hope that many more will benefit from this easy to understand and fundamental book and that they will continue to use it as an on-going resource. I certainly have and our firm is better for it.

Bayardo J. Chamorro
President & CEO
Pribuss Engineering Inc.
South San Francisco, California.

AUTHOR'S FOREWORD

The topics addressed in this textbook respond to the student learning outcomes of the American Council for Construction Education [ACCE]. It is arranged into nineteen chapters in three parts covering contracts, risks and regulations. From 38 to 40 hours of classroom contact hours are recommended to cover this subject matter.

Part I almost exclusively focuses upon the basic elements in the construction contract between the owner and the general contractor; covering this in the textbook's first 10 chapters.

Part II is more expansive in its focus; endeavoring to provide the student with a basic understanding of risk. It covers the following topics:

Chapter 11 - Bidding on Public Projects: Introduces the Federal Acquisition Regulations and bidding processes as a precursor to contract formation. The bidding processes and contractual clauses specified by the FAR have become ubiquitous in the field of public agency procurement. They also inform bid processes, the standard form construction contracts, and general conditions in private contracting.

Chapter 12 - Subcontractor Protections and Vulnerabilities: Discusses the unique contributions of subcontractors and suppliers to the building process; the contractual forms and statutory provisions that govern the relationships between contractors, subcontractors and their suppliers; and the contours of the particular circumstances of subcontractors and suppliers, including the influence of the owner, their relationship with the general contractor, and bargaining disparity. It introduces the flow-thru clause, the pay-when-paid clause, and mechanics' lien laws.

Chapter 13 - Ethical Considerations for Constructors: Explains the elements of a written code of ethics, with particular emphasis upon antitrust law violations. Legal principles are introduced that relate to irrevocable subcontractor bids, bid shopping, bid peddling and reverse bid auctions.

Chapter14 - Construction Insurance: Explains negligence, the common construction insurance products, the legal principles that govern insurance transactions, and the common contractual risk allocations.

Chapter 15 - Surety Bonds: Explains the common construction surety bonds and the risks that contractors assume through indemnification agreements

Foreword

Part III delves into regulatory influences upon the construction industry; including licensing and entitlements, environmental law, labor law, and employment law, as follows:

Chapter 16 - Licensing and Entitlement: Surveys nationwide licensing requirements for contractors. As an illustrative example, it examines statutory proscriptions to contract enforcement in California. It then surveys entitlements including the Davis-Bacon Act, set-aside programs, affirmative action, and the Buy America Act.

Chapter 17 - Environmental Law: Covers Federal and California regulatory schemes for asbestos-containing materials; the Federal radon program; the Federal and California regulatory scheme for lead-based paint; the Federal regulatory scheme for polychlorinated biphenyls; and storm water pollution prevention plans.

Chapter 18 - Labor Law: Recites a brief history of the U.S. labor movement; an overview of the function of the National Labor Relations Board; a description of the unionization process including unfair labor practices; a survey of pressure tactics including strikes, lockouts, and picketing; remedies under the National Labor Relations Act; and a description of union labor agreements.

Chapter 19 - Employment Law: Surveys U.S. employment law topics affecting the construction industry including: the employment-at-will doctrine; employment discrimination; the Equal Pay Act, fair labor standards; the Occupational Safety and Health Act; and labor law enforcement mechanisms.

The intended audience of this textbook is the undergraduate construction management student. The practicing construction management professional will also find this to be a useful reference for rounding out his or her knowledge of construction law. Its real value, however, is less about conveying fundamental factual knowledge than it is about conveying fundamental understandings and in advancing the reader's ability to apply that understanding. The ability to apply an understanding of construction law puts students and professionals alike in the position of being better able to make informed management decisions under the pressure of an active construction project.

PREFACE

Every construction project's field manager can manage jobsite logistics, optimize resource loading, design a critical path schedule and do all of the other technical things that a manager does. These are all technical problems. Managers are paid to solve technical problems. And most managers will perform well at the technical jobs that they are paid to do.

But a construction project is really not a technical problem. The right piece of equipment will not arrive if its supplier decides to send it somewhere else. That planned workforce density becomes irrelevant when a labor strike is called. And you can throw your very logical CPM schedule away when hazardous materials are discovered on site. Thousands of people from hundreds of different companies may come together for a single project. They may have never worked together before. When the project is over they may never work together again. The construction manager has a human relations problem to solve: how to get strangers to cooperate with each other.

It is through the study of construction sciences that we learn how to solve the technical problems posed by construction projects. But the construction sciences are rather silent about human relations problems. They do not solve those problems. These human relations problems degenerate into time-consuming and costly claims and disputes. When people are unwilling or unable to solve their disputes they end up in court.

The legal system does not have the luxury of choosing the problems that it wishes to solve and those that it would rather not. The legal system must solve every problem that comes its way. It has no other choice. That a problem is too messy or intractable does not excuse the legal system from solving it. The messiest and most intractable construction problems are also the most important problems to solve. They tend to engage the most significant and timely issues. If you want to learn about the most significant and timely issues affecting the construction industry, if you want true insight, you must look to the law. The most significant and timely issues are not dealt with anywhere else but within the legal system.

Construction law is a patchwork of common law judicial decisions, statutory laws, and regulations. It is sometimes criticized for being overly complex. It is certainly complex, but not by choice. People are complex. To the extent that the legal system solves people problems, it reflects their complex nature. Construction law has no choice but to reflect the complexity of people and the relationships that they form.

PART I - CONSTRUCTION CONTRACTS

Chapter 1 AXIOMS OF CONSTRUCTION CONTRACT LAW

1.1 Introduction

Few large contractors would start construction without a written contract. Most contractors have a fill-in-the-blank construction contract ready for each new job. The largest contractors retain construction law firms to draft custom contract forms. Medium-size contractors lack the financial resources to pay for

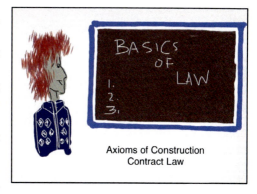

Axioms of Construction Contract Law

custom contract forms. They can, however, develop their own custom contract form by starting with a contract drafted by a lawyer and adding new clauses after each new job, dispute or lawsuit. In time, this contractor has his own custom contract form (Stipanowich, 1998, p. 504). Some small contractors or subcontractors prefer handshake agreements (Stipanowich, 1998, p. 503). They will try to avoid all written contracts. Any contractor, large or small, can rely upon standard form construction contracts such as those published by the American Institute of Architects [AIA], The Engineers Joint Contract Documents Committee [EJCDC], or ConsensusDocs™ endorsed by the Associated General Contractors of America [AGC]. Yet, most contractors will modify these standard form construction contracts to suit their particular needs.

Reviewing a construction contract seems difficult to accomplish. In part this is because every contractor has a different contract form. In part, this is because contracts are so wordy. But if we look through the bewildering array of different and wordy construction contract forms we learn that it is not so much about understanding what is written as it is about understanding what is <u>not</u> written. Few people know how to recognize what is missing in a written contract. The law will fill in the blanks, so to speak, wherever the parties are silent (Staak, 2008). Knowing how the law will fill in the blanks makes all the difference. The commonly heard advice, "Just read the contract," is not very good advice. Better advice is, "Learn the law. Then read the contract."

This chapter attempts to explain important axioms of construction contract law. These axioms infuse meaning and reason into construction contracts although the axioms themselves need not be, and rarely are, expressed in the words of the contract. By learning these axioms we gain a sound, fundamental understanding of construction contract law. This enables us to comprehend the construction contract.

1.2 Meaning and Reason for Contract

A contract consists of promises between people that are enforced by the courts. Their promises must be for legal purposes, however, because contracts for illegal purposes cannot be enforced in any court. Only promises for legal purposes can be enforced. People who want illegal purposes enforced must resort to intimidation and other means – even crime itself – to enforce their contracts. When the courts enforce legal contracts they use civil processes: damage awards, court orders and injunctions. The point here is that intimidation and crime recedes wherever legally enforceable contracts are possible.

You make a promise to build something for a customer who makes a promise to pay you. On your words the court system will enforce both promises. This is a wonderful thing. It means that your word alone is powerful. It means that you can rely on the words of another. It is hard to imagine doing any business today without the power and security that comes from the existence of legally enforceable contracts. Scholars argue that a robust, modern economy depends upon the existence of legally enforceable contracts. In the absence of legally enforceable contracts a modern economy cannot survive.

A contract is a promise. A contract is not like a specification. A specification is an instruction. Everything that you need to know about a specification is written into or referenced by the specification. A specification is always a written document. In contrast, a contract does not have to be written. It can be spoken. It can be spoken but it does not have to be spoken. A contract can be formed by your actions too. It doesn't matter if anything is ever written or spoken about your actions. Your actions alone may be enough to form a contract. And even when a contract is put into writing, many of the things that you need to know about that contract will neither be written into nor referenced by the writing. Federal laws, laws enacted by state and local legislatures, and regulations are effective whether the contract calls them out or not. Particular clauses in your written contract may be invalid for being contrary to those laws and regulations. You are also bound by certain common law duties just because you have entered into a contract.

It doesn't matter if those duties are written in your contract or not. Your duties are implied.

But what forms a contract? What spoken words will form one? What are the consequences of our spoken words? What actions will form a contract? What are the consequences of our actions? We will explore these questions.

1.3 Sources of Law

In order to understand your contract you first have to have a grasp on the laws and regulations that affect your work. After we look at the sources of law we will delve into the most important areas of law that affect construction contracts.

Statutory Law

Laws enacted by federal or state legislatures (congress or senate) are called **statutes**. Laws enacted by municipal and county legislative bodies are called **ordinances** Federal, state and local regulatory agencies enact laws that are called **regulations**.

Most of the laws that we encounter in our daily lives are written by **regulatory agencies** such as the Contractors State Licensing Board, the National Labor Relations Board, and the Environmental Protection Agency. Regulatory agencies are established by statute to establish and enforce standards. Regulatory agencies are semi-independent of the executive branch, subject only to oversight. Regulations have the force of law. Many agencies will also exercise a tribunal function, often with a quasi-judicial official called an administrative law judge. These judges are not part of the court system.

Statutes, ordinances, and regulations are spoken of collectively as **statutory law**. Law can also arise from executive orders of either the President of the United States or the Governors of the various states. Executive orders are also referred to as statutory law.

Statutory laws are grouped together by topics and each group of topics is published as a **code**. Statutory laws of the State of California, for example, are organized into 29 codes. California's 29 codes also incorporate relevant sections of California's Constitution. Codes are given descriptive names such as the Business and Professions Code, the Public Contract Code and the Commercial Code.

Common Law

The judicial branch of government also makes Law. Such law, know as **common law**, consists of the written opinion of judges explaining decisions they have made on cases brought before their courts. Some federal and state regulatory agencies not only write regulations but they can resolve disputes that are brought before them. The written opinions of adjudicators (judges) within regulatory agencies are included in the category of common law.

Court opinions are published in public governmental and private commercial journals, called **reporters**. Each reporter assembles written judicial opinions, known as **cases**. Different reporters cover specific categories of cases. Cases are referenced by **citation**. A citation is a reference system that states the volume and page number where a published case begins in a reporter as bookends to the abbreviated name of that reporter.

In summary, a law can arise from any of the three branches of government and all laws will fall into one of two categories, statutory law or common law. Statutory law is grouped by topic and published in codes. Common law cases are published in reporters.

State Law

There has been a lot of construction law legislation in the various states. Some legislation - known as consumer protection law - aims to protect homeowners. Other legislation, thanks to lobbying by the Associated General Contractors of America [AGC] the Associated Specialty Contractors, Inc. [ASC], the American Subcontractors Association, Inc. [ASA], the Associated Builders and Contractors, Inc. [ABC], and other trade groups, aims to protect general contractors and subcontractors (Sweet, 1997). Laws arise from the various states and each state makes different choices. As a consequence, construction law is different in every state.

1.4 Consideration

Consideration is the substance of a bargain: the promises that are exchanged between the parties (Eisenberg, 2002, §2). Usually, the contractor bargains to earn money while the owner bargains to get a building. Both parties give up something in the transaction too. The contractor gives up labor, material, and equipment, takes on risk and commits time and effort. The owner gives up money. What both parties get and give (in law the words "benefit" and "detriment" are substituted for "get" and "give") is collectively referred to as the consideration (Eisenberg, 2002, §2). For a contract to be lawfully formed, both parties must provide sufficient consideration. The Latin phrase *quid pro quo*, meaning "something for something," provides a good, practical definition for the legal meaning of the word consideration. So for a contract to be lawfully formed, both parties must provide something for something

Donative Promises

Donating a gift is a very kind thing to do, but a promise to donate a gift will not be enforced in common law because gift giving lacks sufficient consideration (*Schnell v. Nell*, 1861). Let's examine a promise to give a gift - what the law calls a **donative promise** - to see how it lacks sufficient consideration. Figure 1.1 illustrates the typical consideration in a construction contract.

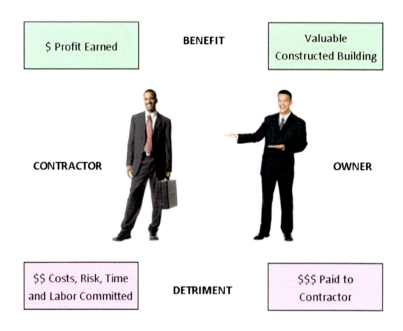

Figure 1.1 Consideration in a Construction Contract

In the typical construction contract the contractor's benefit is the profit he earns. His detriment is the costs, risk, time and effort committed to the construction project. The owner's benefit is the valuable constructed building that he receives. His detriment is the money that he pays to the

CONSTRUCTION CONTRACTS

contractors. This is a *quid pro quo* of sufficient consideration. Four elements are present.

Now let's contrast this with a donative promise. Figure 1.2 illustrates the typical consideration in a donative promise.

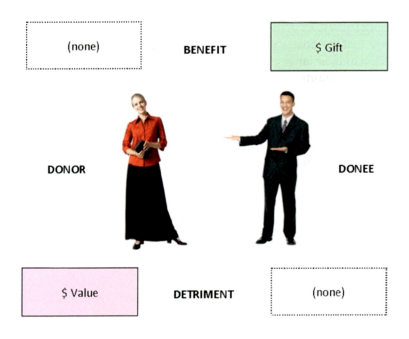

Figure 1.2 Consideration in a Donative Promise

In the typical donative promise the donor receives no benefit. Her detriment is the value of the gift that she is giving up. The donee gets a valuable gift. He suffers no detriment, however. The donor is lacking benefit and the donee is lacking detriment. Because these things are missing, the consideration is said to be insufficient. Perhaps the donor benefits from the feeling of joy that comes from giving while this donee suffers detriment by a feeling of obligation to the donor. Common law courts have reasoned that such feelings are not concrete enough to qualify as elements of consideration. Donative promises are unenforceable because they lack sufficient consideration.

English common law courts established the principle of consideration long ago. It endures today and it is accepted, without serious challenge, as one of the fundamental principles of construction contract law. Perhaps the most important remnant of the long established common law rule of consideration is the court's reasoning on donative promises. First, it was acknowledged that donations were important, that it was in the best interest of the public good for people of means to donate to worthy causes. But it was thought that if promises to donate were enforced in court, people might be reluctant to donate, for fear of a lawsuit. Secondly, even hundreds of years ago lawsuits were backed up in court. As a practical matter it was thought that the courts would be overwhelmed if donative contracts were to be added to their large caseload. These reasons endure today.

Modern courts have found substitutes for consideration that enable contracts to form, in special circumstances, without the need for *quid pro quo*. One of the first such instances where a modern court expanded on the consideration doctrine had to do with bid practices in the State of California. We will explore that circumstance elsewhere, in chapter that examine bidding on public projects.

Bad Bargains

The common law does not care if, when forming a contract, a person gets adequate value in the bargain (Calamari & Perillo, 1999, p. 177). An enforceable contract can be formed as long as both parties receive a benefit while suffering a detriment, even if any of the benefits or detriments is small. Bad bargains still make valid contracts. As long as *quid pro quo* can be found in the bargain, the consideration is sufficient to form a contract. So be careful: sufficient consideration can form an enforceable contract, even if it turns out that that consideration provides inadequate value to one of the parties.

1.4.1 Mutual Assent

The essence of a contract is an exchange of promises. A contract is formed when promises are exchanged in the process of making an offer and its acceptance. Those two words, **offer** and **acceptance** have precise, legal meanings. When trying to determine if a contract was formed the courts can spend a great deal of time examining the offer and the acceptance, measuring the legal meanings of those words against the particular circumstances that occurred. Complicating matters further are other words with precise, legal meanings of their own: **counter-offers**, **rejection**, **revocation**, **expiration**, and many other complications. Exactly what transpires between parties may be examined in detail by the courts for it to be ascertained that a contract was formed or not.

Forming a contract through the process of offer and acceptance also presumes that both parties who exchanged promises possessed **mutual assent** to be bound by their promises. With construction contracts, mutual assent can be more important than offer and acceptance. It is not that offer and acceptance do not matter. It is just that mutual assent is a more significant issue. With construction contracts that are awarded to the low bidder – a bid is the offer and the award is the acceptance – offer and acceptance issues are rather straightforward. Most construction contracts are low bid contracts. Negotiated contracts present themselves differently. While it can be difficult to trace offer and acceptance in the aftermath of a negotiation, it can be easy to show mutual assent as a conclusion to the negotiation. Verbal contracts and documents that might or might not be contracts, agreements to agree are especially problematic. In all of these problematic cases, mutual assent can be a critical issue.

Mutual assent exists when the actions of both people, as would be observed by a reasonable other person, indicate that there was an agreement between them (Sweet & Schneier, 2013, §5.03B). So if a contractor is building something and the owner is making payments to the contractor, a reasonable person would conclude that that contractor must have assented to building, the owner must have assented to paying for what was being built, and there was an agreement between them to do just that. Upon establishing mutual assent it remains only to determine the substance of the promises in order to determine that an enforceable contract had been formed.

Meeting of the Minds

Mutual assent is sometimes characterized as a **meeting of the minds**. The phrase infers that both parties had the same thoughts in mind. But what if they did not have the same thoughts in mind? What if one of the parties had no intention of being bound by a promise; or did not understand the promise and therefore was not of the same mind as the other party; or was joking while the other party was serious?

Older courts would look, subjectively, at what was on the minds of both parties in order to determine if mutual assent was present. If one party thought there was a contract but the other did not, mutual assent to form a contract was not present. There was no meeting of the minds. Modern courts think differently. Modern courts will consider only what a reasonable person could objectively observe (Sweet & Schneier, 2013, §5.03B). If the parties are observed doing things that contracting parties would do, then there is a meeting of the minds. This objective standard often will result in the determination that a contract exists, even if one or both of the parties think otherwise. If they quack, they're ducks, you might say. Unfortunately, courts are creatures of habit. They continue to use the older phrase, meeting of the minds, despite that they attach an entirely different and objective meaning to the phrase (Eisenberg, 2002, §154).

Implied-In-Fact Contracts

Mutual assent, sufficient to form a contract, can occur even without spoken words of agreement (*U.S. v. Young*, 1974). Such contracts are known as **implied-in-fact contracts**. Here is an example. A ready-mix truck filled with nine yards of concrete gets stuck in the mud while struggling uphill on a muddy jobsite. The driver waives his arms at a crane operator who lowers his hook. The driver then rigs a cable between the hook and his truck, enabling the operator to winch the truck out of the mud. Later, the ready-mix company gets a bill for use of the crane and the operator's time. Do they have to pay? The answer is yes. As long as their bill is for a reasonable amount of money – easily determined if the crane company has a published schedule of hourly billing rates – the ready-mix company will have to pay it. This is an implied-in-fact contract and the courts will enforce it.

The crane company provides a service for profit and the concrete company provides materials for profit. A reasonable person would look at the actions of the driver and the operator and conclude that they reached a mutual agreement to provide specific services in the course of conducting their respective businesses. Neither company is in the business of doing things for free. So if the rate charged by the crane company was reasonable, you have an enforceable, implied-in-fact contract. This is a real contract. Unlike quasi-contract and unjust enrichment for which the only remedy is restitution, the remedy for an implied-in-fact contract is expectation damages, in other words, costs <u>and</u> profit (Sweet & Schneier, 2013, §156) (Farnsworth, 2011, §4).

Bilateral and Unilateral Agreements

Contracts involving the exchange of one's promise for another's promise are called **bilateral agreements**. *Quid pro quo,* in the construction context, usually involves the exchange of promises. But a contract could also have one party exchanging a promise for the other party's action. The implied-in-fact concrete truck example was one example of such an exchange. In that example the driver exchanged his implied promise to pay for the operators' actual act of swinging his hook over to rescue the driver's truck. Unlike a bilateral agreement, this implied contract did not form on the exchange of promises. It formed only when the operator actually swung his hook over. Contracts exchanging a promise for an act are called **unilateral agreements**. Both bilateral and unilateral agreements are enforceable as long as the consideration is sufficient (Eisenberg, 2002, §§ 6-7).

1.4.2 Verbal Contracts

The essence of a construction contract is an exchange of promises between two people: one person promising to build something and the other promising to pay for what will be built. The law enforces these promises. Should either person break his promise the law finds a remedy.

Typically, the promise to build and the promise to pay are exchanged as spoken words. Those words may be all that it takes to form a legal contract. Later, those words may be put into writing, both parties attaching their signatures. What results is a written construction contract. The process of putting spoken promises into writing is called **memorializing** an agreement (Sweet & Schneier, 2013, §5.05A). Keep it in mind, however, that neither the owner nor the contractor need ever put their words in writing just to form a legal contract. With a few exceptions that we will touch on later, verbal promises exchanged between people can form real contracts, fully enforceable by the courts (Sweet & Schneier, 2013, §5.06A).

To enforce a verbal contract, the spoken words must be sufficiently clear to define the promises (Sweet & Schneier, 2013, §5.03D). Promises can be defined with very few words as long as the meaning of those words is clear. It would not do, for example, for a contractor to say, "I will build a house for a good price." Those words are too vague because they do not explain how big a house, how many rooms, what materials, the quality of construction, or what it will cost. But words like "I will build a house for you just like the one across the street, for $150,000" are clear. Those words provide two measurable standards: 1) the house across the street, and 2) $150,000. Those standards are enough to enable a reasonable person to define both the promise to build and the promise to pay.

There is more to forming a construction contract than defining the promises. But for now, let us just focus our attention on defining the promises. Other problems in contract formation will be covered elsewhere.

The usual problem involving a verbal construction contract has one person asserting that the contract exists while the other is denying that the contract exists (*Meco v. Dancing Bear*, 2001). Another common problem occurs when a verbal construction contract is put into writing but one of the parties does not sign. Particularly if the work had started and the contractor has received partial-payment, the court will likely find that there was a verbal contract (Sweet & Schneier, 2013, §5.05A). As long as the court has some way to define the substance of the promise that was broken a verbal contract can be enforced.

The Statute of Frauds

The phrase **statute of frauds** is a misnomer because it is not a statute in the usual sense but a rule of common law requiring certain types of contracts to be in writing. The types of contracts that must be in writing - by the common law statute of frauds - include a contract for the sale of land, a contract for the sale of goods over $500, or a construction contract that is impossible on its terms to be completed within one year.

The phrase "impossible on its terms to be completed within one year," is frequently misinterpreted to mean any construction work that takes longer than one year to complete. But the statute of frauds applies only to contracts that are literally impossible to complete within one year of the date that the contract is executed (Calamari & Perillo, 1999, p. 257). Nearly any construction contract can be completed in less than a year if you throw enough money at the work. Most owners would find it difficult or impractical to do that but the statute is not concerned with how difficult or impractical it is to complete the work within a year. It must literally be impossible to complete within one year for the statute to apply. The statute would apply, for example, to a contract where the work cannot start until a year-and-a-day after the contract is signed. That contract is "impossible on its terms" to be completed within one year; therefore, it must be in writing. Very few construction contracts fit this rule.

1.5 Remedies

When money is lost because a promise was not fulfilled, the law characterizes the person who lost the money as having been injured. The law attempts to enforce construction contracts by putting injured parties into the financial position that they would have been in had the other party's promise been fulfilled (Dobbs, 1993, §12.1(1)).

Damages

When a contractor breaks his promise by not completing work, the court's usual remedy is to award to the owner the amount of money it takes to have the job completed by another contractor (Dobbs, 1993, §12.19(1)). When an owner breaks his promise by not paying a contractor, the court's usual remedy is to enable the contractor to recover an amount of money to cover his costs, an appropriate share of his overhead costs, and the profit that he expected to earn had the job been completed (Dobbs, 1993, §12.20(1)). The amount of money awarded is the measure of **damages**. Damages awarded to contractors are sometimes called **expectation damages** because the award includes the profit that the contractor expected to make.

Courts have the power to order, or enjoin, a party to a construction contract to do something. Such orders are a type of remedy referred to as **specific performance**. Courts may also order a party to refrain from doing something. An order to refrain from doing something is accomplished through an **injunction** (Sweet & Schneier, 2013, §5.10A).

Courts will always seek to award damages as a remedy before resorting to specific performance or injunction (Sweet & Schneier, 2013, §5.10B). They will order specific performance or injunction only when damages are shown to be an inadequate remedy or when actual damages are difficult or impossible to determine. Let's say that a contractor installs the wrong brand of pipe. If that pipe is a code violation, especially if it exposes the homeowner to harm, the court is likely to invoke specific performance, ordering the pipe removed and replaced with the specified pipe.

Yet if that contractor had installed a non-specified brand of pipe that was not a code violation – it was just a different, perhaps cheaper brand of pipe – the court is not likely to order it ripped out and replaced with the specified brand of pipe. To do that would be wasteful. The courts will always seek to avoid **economic waste**. In this circumstance, the court is more likely to award the owner with damages equal to the loss in market value, if any, of a house built with the wrong brand of pipe (*Jacobs & Youngs v. Kent*, 1921).

Restitution

But what happens, you might ask, if the court cannot figure out what was promised? In that event, a contract cannot be enforced because the substance of the promise is too vague. All is not lost, however, for a contractor who finds himself in such a dilemma. Let's say, for example, that a developer and a contractor make a verbal contract to build a house. The contractor builds that house at a cost of $250,000. The developer then refuses to pay the contractor. The contractor sues in breach of contract. The court finds that there probably was a verbal contract between contractor and developer because the developer knew that the contractor was building for him and contractors don't work for free. Despite the apparent verbal contract, the contractors promise was so vague that the verbal contract could not be enforced. Does this contractor have a remedy by which to recover the $250,000 that he spent on labor, material and equipment? The answer is yes; the law has a remedy for this situation. This remedy is called restitution.

Restitution enables a contractor to recover his actual costs but not any of the profit he expected to make (Dobbs, 1993, §12.11(1)). In contrast, damages enable a contractor to recover his actual costs plus expected profit. The difference between the two is that damages include profit while restitution does not. Restitution is sometimes called a **quasi-contract** action. This phrase is a misnomer though, because restitution is not a contract. Restitution is the remedy of choice when no contract exists between the parties.

Restitution is also the remedy of choice for an injured contractor whose total construction costs exceed its contract price – a losing contract (*Bauman, York & Bauman*, 2003). If a contractor with a losing contract were to choose the usual formula for damages – cost plus profit – with profit plugged into the formula as a negative number, the amount of money that it recovers under the formula would be less than its costs. Fortunately, injured parties are able to choose restitution instead of damages (Sweet & Schneier, 2013, §12.7(6)). Having a losing contract, a sensible but injured contractor will choose restitution over damages.

Unjust Enrichment

Restitution can also be invoked in the infrequent and unfortunate circumstance of a building partially or entirely upon the wrong property (*McCreary v. Shields*, 1952) (*Voss v. Forgue*, 1956) (*Beacon v. Holt*, 1966) (*Somerville v. Jacobs*, 1969). Here is a situation where there may be a perfectly valid but irrelevant contract between the contractor and someone else. The contractor has no contract with the property owner upon whose property his just completed house now happens to be standing. This contractor has two big headaches, recovering his costs for building the house on the wrong property and covering the cost of building another house on the right property.

Restitution arises from the doctrine of **unjust enrichment**. Simply put, unjust enrichment means that it would be unjust for a homeowner to receive a house for free at the expense of a contractor who did not intend it to be free. One remedy for this particular injustice would be to disgorge the homeowner of the contractor's cost to construct the house, but not the contractor's profit. Another remedy would be to disgorge the owner of the house's market value. These remedies could be invoked here from the doctrine of unjust enrichment as long as the property owner had no knowledge of the construction and the contractor did not know of his error until it was too late.

When a contractor builds on the wrong property, the court can force the sale of some or all of the innocent property owner's property to the contractor at fair market value. There are also legal remedies other than a forced-sale by which the court may enable a contractor to someday recover some or all of his costs. As long as both parties are innocently involved in the mistake, the court will try to take some of the sting out of the contractor's dilemma but not so much for him to profit from his mistake and without unduly penalizing the innocent property owner.

Quantum Meruit

The Latin phrase **quantum meruit** means "as much as he deserves." Courts will sometimes fashion a remedy in *quantum meruit* when an enforceable contract between two parties does not exist but the court finds that a contract was implied in law. *Quantum meruit* differs from unjust enrichment in that a contractor can recover actual costs plus reasonable profit and overhead with *quantum meruit*. Unjust enrichment has the remedy of restitution under which a contractor can recover its actual costs but not any profit or overhead.

The most common application of *quantum meruit* occurs when the parties to an enforceable contract abandon that contract but work continues thereafter. Abandonment requires both parties to abandon their contract. Abandonment cannot arise from the actions of one of the parties alone. Such circumstances might arise when major changes to a project or its schedule are so severe that the continuing work no longer resembles what was originally contemplated by the parties in their contract but the contractor continues working anyway and the owner continues paying the contractor's invoices. Such extreme changes are known as cardinal changes.

For the courts to apply *quantum meruit* in a construction setting the following elements must be in place:

1. An enforceable agreement between the parties was absent (or had been abandoned);
2. Work must have been performed by the contractor in good faith;
3. The owner must accept the work performed by the contractor;
4. There was an expectation of reasonable compensation for the work that was performed; and
5. There exists a means for determining the reasonable value of the work that was performed.

1.6 Agreements to Agree

In an ideal world, contractor and owner would exchange promises, their promises would be sufficiently clear to be enforceable and they would memorialize their promises in a signed, written contract. Business in the real world is not that simple. Sometimes an owner will want a contractor to start work before they have reached agreement on a final, written contract. In order to get started they will make an **agreement to agree**. The key legal question is whether their agreement to agree is an enforceable contract or not.

Agreements to agree are labeled with various titles. Some of the most commonly used titles are **letter of intent** [LOI], **memorandum of understanding** [MOU], **temporary contract**, **interim contract**, and **preliminary contract**.

Letter of Intent

A **letter of intent** is a written instrument that directs a contractor to start work before having reached agreement on a final, written contract. Work will commence on many projects under an LOI while the final contract is still being negotiated (Sweet & Schneier, 2013, §5.07B). In-house design, engineering, management time and other salaried expenses, so called **soft costs**, are easily committed without a written construction contract. Purchases of material, trade labor, equipment, and subcontracts, so-called **hard costs**, are more difficult and risky to commit. Nevertheless, it is not unheard of to commence and even to complete work despite never coming to an agreement on the final contract.

For an LOI to have meaning it must include a formula that describes how and when payment will be made. That way, if the parties do not come to an agreement their disputes can be settled without going to court. If a dispute must be settled in court, the courts will examine the words of the LOI and the circumstances that gave rise to the LOI in order to determine if a contract was or was not formed (Sweet & Schneier, 2013, §5.07B). A contractor who is proceeding on an LOI is at risk of losing profits but is likely to recover some or all of its costs through restitution. Many contractors are willing to take that risk over soft costs but unwilling to take that risk over hard costs.

Usually an LOI will limit performance, either by specifying a period of time, such as "for the next 30 days" or a category of work, such as "site work and excavation." The courts will often rule that these are enforceable contracts, especially if work has actually been performed and the contractor has received some amount of payment.

Memorandum of Understanding

Companies will write a **memorandum of understanding** to summarize what has and has not been agreed to while negotiating with the intent of forming a contract. Generally, an MOU does not form an enforceable contract (Sweet & Schneier, 2013, §5.07C). Among other enforceability problems, terms that are settled in a MOU can become unsettled at any time prior to concluding negotiations. Exactly what has been promised remains in flux during negotiations: promises do not become firm until the negotiation is over. Whenever promises are not clear, the court will not enforce them as a contract. Therefore, an MOU is generally not sufficient to form an enforceable contract.

Inchoate Contracts

Yet another type of agreement to agree is an **inchoate contract**. An inchoate (pronounced in-ko-it) contract is a partially completed or imperfectly formed contract (Garner, 2006). You will find inchoate contracts labeled with the titles **temporary contract**, **interim contract** or **preliminary contract** (Sweet & Schneier, 2013, §5.07E). In common practice these phrases are meant to describe an incomplete contract where the intent is to replace it with a complete contract at a later date. Unfortunately, the words "temporary," "interim," and "preliminary" lack precise, legal meanings. The courts will determine whether or not they form enforceable contracts on a case-by-case basis. In general, they will look to the wording of the inchoate contract, the facts that surrounded its formation and the circumstances of any dispute in order to determine if it creates an enforceable contract or not.

Enforceability of Agreements to Agree

In the past, courts would not enforce an agreement to agree. Modern courts are more inclined to enforce these as contracts (Sweet & Schneier, 2013, §5.07A). If work has started and the clauses that aren't yet agreed to are not critical to the determination of damages, they will enforce the agreement to agree. Sometimes the court can fill in the blanks of the not-yet-agreed-to terms with what is found in usage of trade. Usage of trade is any customary, local construction method or practice. By filling in the blanks in this way, the court might be able to determine the substance of a contract.

That enables the court to enforce an agreement to agree. In general, courts will look to the wording of an agreement to agree, the facts that surrounded its formation and the circumstances of any dispute in order to determine if it creates an enforceable contract or not.

1.7 Implied Warranty

A **warranty** is an assertion that what is being provided is as promised. The warranty obligation arises either from the express words in a construction contract or it arises out of the circumstances of construction work, not because of any expression that is or is not in the construction contract. A warranty that arises from the words in a construction contract is called an **express warranty**. A warranty that does not arise out of the words in a construction contract is called an **implied warranty**. Both types of warranties have the same effect in that they create obligations for which the contractor will be held liable for any failure to perform.

Workmanlike Performance

Most courts will find a warranty of workmanlike performance implied in construction contracts (Davis, 1993) (*Scott v. Strickland*, 1984) (*Lempke v. Dagenais*, 1988). **Workmanlike performance** is a legal maxim. In essence, it means that when a contractor undertakes to perform work he brings the requisite skill of his profession and if he does not, and fails, he is liable for the consequences (Davis, 1993). Because workmanlike performance is implied in all construction contracts a contractor is obligated to provide workmanlike performance whether the words of his contract express that obligation or not.

Most people understand a warranty to have effect when something is found defective after construction is complete. It is that. But a warranty can also have effect during construction. The implied warranty of workmanlike performance is such a warranty. A contractor who persistently fails to supply suitably skilled workers could suffer the consequences of breaching his implied warranty of workmanlike performance. The consequence of that breach could be early termination of the contractor's contract plus the award of damages to the owner. Damages in this circumstance are usually in the amount of any difference in price for having a different contractor finish the work.

Warranty of Fitness

A **warranty of fitness** is an implied warranty that asserts that the work, once completed, will be fit for the ordinary purpose of the work. The implied warranty of fitness also asserts that the work, once completed, will be fit for any special purpose of the work provided that the contractor knows, or should know, what the special purpose of the work is. The implied warranty of fitness can be changed by the express words in the contract, however.

That being said: as long as a contractor builds to the plans and specifications he will not be held liable under the contractor's implied warranty of fitness (*Sunbeam v. Fisci, 1969*) (*Corporation of Presiding Bishop v. Cavanaugh*, 1963) (*Mannix v. Tryon*, 1907). But if a contractor does not build to the plans and specifications he will be held liable. The contractor also provides a warranty of fitness when the owner provides neither plans nor specifications at all to the contractor (*Aced v. Hobbs-Sesack*, 1961).

In California, all construction contracts have an implied warranty of fitness (*Pollard v. Saxe & Yolles*, 1974). All such implied warranties are enforceable by the original owner of newly built property. California has extended their implied warranty of fitness beyond original owners, to subsequent owners of newly built property (*Pollard v. Saxe & Yolles*, 1974). So far, however, this extension has been limited to commercial construction and it has not been applied to the construction of single-family homes (*Siders v. Schloo*, 1987).

1.8 Good Faith and Fair Dealing

Before a contract is formed between them, the contracting parties are expected to look out only for themselves (Sweet, 1997, §20.04). The courts generally do not expect them to look out for each other (Sweet, 1997, §20.04). Once a contract is formed, however, that "care not" posture changes. This is especially true when a construction contract is formed.

Most contracts are about a single transaction where one party is selling what the other party is buying. This type of contract can be called a **transactional contract** because it has to do with a fixed transaction. Construction contracts are also about buying and selling. However, unlike a transactional contract, the substance of the contract is likely to change before the contract is completed. A construction contract can be called a **relational contract** because it attempts to describe the future relationship between the contracting parties as the substance of their contract changes (Stipanowich, 1996). Modern construction projects cannot avoid change because these projects are very complex. Such complexity cannot be captured with certainty in any set of construction documents. Change is unavoidable and construction contracts must reflect this reality.

Modern courts have found that certain duties are implied in relational contracts, whether or not those duties are expressed in the words of the contract. One of the most important implied duties is the duty of **good faith and fair dealing** (Sweet & Schneier, 2009, §20.04). Some of the more significant applications of good faith and fair dealing are found in uncovering design problems, the duty of cooperation, and in the reporting of errors or omissions in the design.

Uncovering Design Problems

A contractor must bring design problems to the attention of the architect (*American & Foreign v. Bolt*, 1997)(*Eichberger v. Folliard*, 1988). The courts do not expect contractors to have the same professional competencies as architects and engineers. Accordingly, contractors are not expected to recognize and report each and every potential problem with the design. But they do expect contractors to be competent enough to recognize obvious flaws in the design. A building code violation is a good example of a design problem that a contractor should recognize. Contractors have a good faith duty to bring any design problems that they uncover, or should uncover, to the attention of the architect and/or engineer. If they do not they may be liable for the consequences. Those consequences could include money damages.

Duty of Cooperation

Courts have found an implied **duty of cooperation** in construction contracts (*George A. Fuller v. U.S.*, 1947). An owner cannot interfere with or cause delay to the work of his contractor (*Lewis-Nicholson v. U.S.*, 1977)(*Snyder v. State*, 1960). Contractors are expected to cooperate with the owner, the architect, engineer, other consulting professionals, other contractors, subcontractors and suppliers (*U.S. f/b/o Wallace v. Flintco*, 1988)(*Allied v. Dick*, 1995)(*Crawford v. Bateson*, 1988). A contractor cannot interfere with or hurt the ability of others to perform on their contracts. More importantly, the cooperation of a contractor includes the exercise of skill and judgment by the contractor in order to further the best interests of the owner.

Reporting Errors or Omissions

The courts have found an implied duty of good faith and fair dealing in all construction contracts. Among other consequences, this duty compels a contractor to notify the architect when any errors or omissions are found in the drawings, specifications and other contract documents. The contractor is not expected to have the knowledge of a registered architect or professional engineer. The contractor is expected to have sufficient knowledge, however, to recognize obvious errors or omissions.

Knowing has special meaning in law. It means either actually knowing or knowing what a competent contractor should have known. The ramification of this is that a contractor may incur liability for not having reporting errors or omissions that he should have discovered regardless of whether he actually discovered them or not.

Chapter 2   WRITTEN CONTRACT FORM & SUBSTANCE

2.1   Introduction

The AIA, the EJCDC, and ConsensusDocs™ publish the standard form construction contracts that are most commonly encountered in the United States.

The AIA documents are a suite of standard form construction contracts, addressing all aspects of construction projects. AIA documents have long been the most widely used standard form construction contracts.

Written Contract
Form & Substance

ConsensusDocs™ are a different suite of standard form construction contracts that were developed by consensus of various industry groups representing owners, contractors and designers. ConsensusDocs™ evolved from what used to be the Associated General Contractors [AGC] suite of standard form construction contracts. Both AIA documents and ConsensusDocs™ are focused on commercial design and construction although they are generally suitable for residential construction, particularly large residential developments.

The EJCDC has yet another suite of standard form construction contracts. The EJCDC documents, however, are oriented more toward industrial, heavy civil and highway design and construction than they are commercial or residential construction.

Most contractors will modify the standard form construction contracts for their own purposes. Some will write their own, unique contracts. Every contractor seems to prefer different words in their written construction contracts. But despite that their words will be different, we still recognize the same key contracting issues in each construction contract: responsibilities of the owner, contractor, subcontractors and designer; payments and completion; concealed or unknown conditions; changes; time; interpretation of contract documents; suspension and termination; assurance; claims and dispute resolution; insurance; and surety bonds. These issues are common to all well written and complete construction contracts. To understand the key contracting issues is to understand the meaning of a written construction contract.

The emphasis of this chapter is on the form and substance of the written construction contracts that will ordinarily be encountered in residential construction, commercial construction, heavy civil construction, and alternate project delivery methodologies.

2.2 The Fully Integrated Agreement

Forming a construction contract is a two-part process. The first part of the process usually concludes with a verbal contract. The verbal contract could come into being upon notification of award to the successful bidder or it could come with a letter of intent at the conclusion of contract negotiations. Although there could be some form of writing by either or both parties or perhaps a series of writings emerging from negotiations between the parties, these writings do not rise to the level of a written contract. The second part usually concludes later - sometimes much later - with a written contract. Typically, a written construction contract is a lengthy, formal document that is signed by both parties.

A properly written construction contract will contain a clause that invokes the phrase "…entire and integrated agreement…" This is known as the **merger clause**. The merger clause asserts that the written contract is the complete agreement between the parties. As long as the written contract addresses everything that would normally be included in a contract of the same sort, a written contract with a proper merger clause is considered a **fully integrated agreement** (Calamari & Perillo, 1999, pp. 236-237).

The fully integrated, written construction contract stands alone. The courts apply the **parol evidence rule** to fully integrated agreements. This rule says that the meaning of the contract can be found within the four-corners of the paper and no prior verbal or written contracts can be introduced as evidence to interpret the meaning of the contract. Upon signing a fully integrated agreement, the written contract also gains independent legal significance whereby any prior verbal or written agreements between the parties are no longer of any importance. They are not part of the newly written contract.

When courts attempt to interpret the meaning of partially integrated agreements, they have to look to past verbal or written contracts to determine the original intent of the parties. In the absence of a proper merger clause, courts are divided on whether or not prior verbal or written agreements are a part of the contract. And if the written contract is either silent about something that would normally be included or it lacks a merger clause, the court may consider it to be a **partially integrated agreement** (Calamari & Perillo, 1999, pp. 236-237). Such interpretations are much more complicated, often yielding results that neither party intended.

The standard form construction contracts provided by the AIA, the EJCDC or ConsensusDocs™, and others, have merger clauses and cover everything that would normally be included in a construction contract. Therefore, one would normally expect the courts to construe any of these standard form construction contracts as a fully integrated agreement. This is a very compelling reason for using a standard form construction contract.

The Contract Documents

A written construction contract is a collection of documents. This collection is known as the **contract documents**. The contract documents will include parts that are titled **agreement, conditions, drawings, specifications,** and **additional documents**. All of the contract documents, taken together in their entirety, are the written construction contract. The law interprets the contract documents as a whole, each part providing meaning to the whole.

The owner-contractor agreement is one part in the contract document collection. In it you find:

- the date of the agreement;
- the name of the project;
- the identities of the owner, owner's representative, contractor, contractor's representative and the designer;
- the contract amount;
- unit prices and allowances, if any;
- liquidated damages for late completion, if any;
- progress payment, retention, and final payment conditions;
- the date of commencement of the work;
- the date of substantial completion; and
- a place for the owner and the contractor to sign.

The owner-contractor agreement introduces the work of the contract and describes the contractor's duties to provide labor, materials, equipment, insurance, bonds, and all other services that are required by the contract

CONSTRUCTION CONTRACTS

documents. It then defines the work of the contract by enumerating each and every contract document by number and title, or it references a specification that enumerates them. In this text, the term owner-contractor agreement refers to the written construction contract, the contract documents, or to the agreement part in the contract documents collection depending upon the context in which the term is being used.

The **conditions** are a lengthy part comprised of instructions, information, suggestions, and things that the parties are supposed to do. The conditions constitute the rules-of-the-game; rules that enable the parties to manage their relationship effectively. Embedded within the conditions you will also find legal conditions. A **legal condition** is a provision that has to be fulfilled before a contractual duty arises.

The standard form contracts of the AIA, the EJCDC and ConsensusDocs™ include very comprehensive (and lengthy) conditions. Known as **general conditions** these standardized forms are very popular. Because they get used over and over from project to project, they have become very familiar to owners, constructors and designers alike.

- Title page
- Table of contents
- Addenda (if bound)
- Bidding requirements
- Invitation to bid
- Instructions to bidders
- General conditions
- Supplementary conditions
- Schedule of drawings
- Technical specifications
  - (CSI Divisions 02 thru 49)
- Sample forms
  - Agreement
  - Bid form
  - Bid bond
  - Performance and payment bonds
  - Other sample forms

Figure 2.1 Project Manual Contents

Specifications are written descriptions of the work of the contract. The term specification is used very broadly to include all contract documents with the exception of drawings. Some specifications convey procurement and contracting requirements while others convey technical information. Specifications are ordinarily bound into a **project manual**. Figure 2.1 displays typical contents of a project manual.

Specifications are often configured to the Construction Specifications Institute [CSI] **MasterFormat™**, a widely used construction document classification system. Prior to 2004, MasterFormat™ consisted of 16 divisions. In 2004 it was re-organized into its current 50 divisions.

Bid documents such as an **invitation for bids [IFB]** or **request for proposals [RFP]** and a contractor's bid or proposal and any addenda thereof may be listed as **additional documents**. An IFB is a collection of documents such as bid forms and bidding instructions, typically for soliciting sealed, competitive bids for construction at a firm-fixed price. An RFP is a collection of documents, including bidding instructions, typically for soliciting proposals as a precursor to contract negotiations between the owner and the contractor. **Addenda** are any changes to an IFB or RFP that were issued prior to contract award.

Most owner-contractor agreements will exclude all bid documents from the contract documents with deference to the **parol evidence rule**, the legal maxim of contract interpretation that excludes evidence of prior written or verbal agreements. So you have to be careful here. Bid documents enumerated as additional documents may inadvertently become incorporated into the contract documents. Conflicts between these bid documents and the other contract documents are problematic. Some public entities expect the bid documents to be included in the contract documents. Bid documents should only be enumerated if they are intended to be a part of the contract documents.

More often than not though, bid documents are not intended to be a part of the contract documents. If the bid documents are formatted as project specifications in the CSI MasterFormat™ 2004 system, they all belong to the division 00 procurement and contracting subgroup. When the bid documents are formatted this way it becomes a simple matter to exclude the project manual's division 00 subgroup 1 from the contract documents. Division 00 is one of six CSI subgroups listed in Figure 2.2. Divisions 02 thru 49 are the **technical specifications**. Each technical specification covers the work of a particular trade.

Modifications are change orders that alter the work of the contract and the price thereof that occurs after the date of the agreement. Although valid change orders stand alone as contract documents, some contracting parties prefer to produce a list that summarizes all modification to their agreement. The most convenient place to put this list is in the owner-contractor agreement.

CONSTRUCTION CONTRACTS

Division 00	Procurement and Contracting
Division 01	General Requirements
Division 02-19	Facility Construction
Division 20-29	Facility Services
Division 30-39	Site and Infrastructure
Division 40-59	Process Equipment

Figure 2.2 CSI MasterFormat™ Subgroups

The Work of the Contract

The owner-contractor agreement obligates the contractor to provide labor, materials equipment, insurance, bonds, and all other services that are required by the contract documents. This agreement is between the owner and contractor and only the owner and contractor. Whenever two-parties form a contract they are legally described as being in **privity** or **privity of contract**.

The project may be bigger that the work defined in the owner-contractor agreement. It may involve work by other contractors or work done by the owner. So one must be careful when reading the contract documents because the project may include work that is not within the scope of work of the owner-contractor agreement.

The designer is not a party to the owner-contractor agreement, therefor has no privity of contract with the contractor. A lack of privity does not, however, extinguish the designer's right to enforce certain obligations against the contractor such as: the contractor's duty to indemnify the architect (i.e. to purchase insurance) for certain risks; the contractor's duty to perform on warranty obligations; and the contractor's obligation to allow the designer access to the work.

Insurance and bonds are just as much a part of the work as the bricks and mortar of a building. The contractor is obliged to purchase and maintain insurance and provide bonds. The owner-contractor agreement will include a description of the types, terms, and conditions of the various insurance policies and bonds that are required and the limits of liability or bond amount for each.

2.3 Roles and Responsibilities

An **owner** might be a private individual, a commercial business entity, or a federal, state or local government or government agency. The term **contractor** is interchangeable with the terms "general contractor" and "prime contractor." It refers to a business entity, typically holding a general contractor's license, which enters into a written contract for construction with an owner. Among its other stipulations, the written contract between an owner and a contractor will define their mutual responsibilities.

The term **subcontractor** is interchangeable with the terms "specialty contractor," "trade contractor" or "prime trade contractor." It refers to a business entity, typically holding a specialty contractor's license, which enters into a written contract for construction with a contractor. Subcontractors typically perform construction services aligned with the construction trades: earthwork, ironwork, carpentry, masonry, mechanical, electrical, plumbing and such. Subcontractors might form a contract with other subcontractors, who are called "sub-subcontractors" and those sub-subcontractors may subcontract with "sub-sub-contractor" and so forth. There may be many, many such layers, or "tiers," in a project. The word subcontractor, as is used in this chapter, means the subcontractor, sub-subcontractor, sub-sub-contractor, and others in any tier.

The written construction contract between an owner and a contractor will describe the responsibilities of subcontractors even though the subcontractors are not parties to the owner-contractor agreement. Subcontractor's responsibilities are defined because they deliver various services during construction on behalf of the contractor. The word **constructor** is used throughout this text to mean contractor, subcontractor, or both. It is not used in this particular chapter so as to avoid confusion among terms.

The word **designer** indicates either an "architect" or an "engineer," as the case may be. Either the architect takes the lead role in design or the engineer takes the lead role in design depending on whether the project is a built artifact or an engineered facility. Architects design residential homes, commercial and multi-use low-rise, mid-rise and high-rise buildings, convention centers, sports facilities, hospitals, schools and universities, prisons, institutional buildings, and other built artifacts. A wide range of consultants in specialized fields assists them during the course of design. An engineer designs roads and highways, railroads, bridges, waterways, dams, sewers and wastewater facilities, power plants, industrial process plants and other types of engineered facilities. A wide range of consultants in specialized fields assists them during the course of design. Both lead architects and lead engineers must possess legal registration in the state where the project that they design will be constructed.

The written construction contract between an owner and a contractor will also describe the responsibilities of the designer even though the designer is not a party to the owner-contractor agreement. The designer's responsibilities are defined because the designer delivers various services during construction on behalf of the owner.

The Owner

The **owner** is the person or legal entity that you are contracting with. The owner may or may not actually own the construction site or what will be constructed on that site. There may not even be a site, as is the case with air rights over an existing structure, or a condominium where one has the right to build within space defined by walls without actually owning the walls. A tenant or lessor may own rights to build. Owner infers a person or entity that "owns" the legal right to build what is contemplated in the contract documents.

The owner must have legal authority to sign the construction contract. When the owner is a business entity, its signatory is an officer or agent of that company. When the owner is a government or public agency, its signatory is a contracting officer who is vested with the appropriate authority. This signatory owner is not necessarily the same as the owner with the legal right to build. Regardless, owners will rarely manage the day-to-day activities of a project. Accordingly, the owner designates a representative in writing. All things considered, the "owner" means the person or entity with the legal right to build; and/or the person, officer, or agent who signed the construction contract; and/or a representative designated to act on their behalf.

As a practical matter, the **owner's representative** will be the decision-maker and that's who the contractor's representative will work with. Also important, nothing in the contract prevents two or more owner's representatives from being named. When that occurs, the roles and responsibilities of each should be clearly delineated. The designer is not the owner's representative. However, the designer will have certain duties that mimic the roles and responsibilities of the owner's representative.

The owner must provide the correct legal description of the property. This description must be precise in order enable the courts to enforce a mechanics' lien. Both owner and contractor have the ability to furnish a survey of the site but usually the owner will provide it. This is reasonable because owners usually obtain a survey of the site during the design phase and it would be wasteful for a contractor to spend money to obtain another.

A contractor may want the right to obtain evidence that the owner is able to fulfill its obligation to pay the contractor. Loan commitment letters from private lenders and appropriation commitment letters from public agencies generally provide the needed evidence.

Construction by Owner or Separate Contractors

The owner might contract with many different contractors to perform different parts of a single project. The owner might also use its own employees to perform some of the work. There may also be other construction operations on site that are associated with different projects. Contractors will want to file claims for delays or extra costs when impeded by other construction work. The owner will be responsible for coordinating the contractor's work with the work of its other contractors. The owner may delegate these responsibilities to others. The contractor is required to cooperate with these coordination efforts, particularly with respect to schedules.

Wherever a contractor's work is dependent on the work of other contractors, rules must be established for handing off that work. A contractor is expected to inspect the other contractors' work and report any apparent discrepancies or defects to the designer. When a contractor starts performing work that depends on others, the contractor accepts their work except for the obvious defects that had been reported to the designer. The other contractors remain liable for any discrepancies or defects that were not reasonably discoverable by the contractor's inspection.

Few, if any, of the other contractors on site will have contracts with each other. So if one damages the work of another, the contractor who is harmed will look to its owner-contractor agreement for compensation. The owner will, in turn, look for compensation in its contract with the contractor who caused the harm.

The Contractor

Just like the owner, the contractor must have legal authority to sign the owner-contractor agreement. When the contractor is a business entity, its signatory is an officer or agent of that company. This signatory will rarely manage the day-to-day activities of a project. Accordingly, the contractor designates a representative in writing. "Contractor" means the person, officer, or agent who signed the agreement; and/or a representative designated to act on their behalf. As a practical matter, the contractor's representative will be the decision-maker and that's who the owner's representative will work with. Also important, nothing in the contract prevents two or more contractor's representatives from being named. When that occurs, the roles and responsibilities of each should be clearly delineated.

The contractor must be licensed in locations where licensing is required. In jurisdictions with licensing laws, the courts will not enforce the construction contracts of unlicensed contractors. Nor can an unlicensed contractor obtain a mechanics' lien. In its combined effect this means that an owner can lawfully refuse to pay an unlicensed contractor for any work that has been performed.

The owner usually pays for zoning, environmental impact, and other approvals and permits relating to project feasibility. The contractor usually pays for all permits, fees, licenses and inspections that are customarily secured after execution of the contract. The contractor is responsible for giving notices to public officials for the purposes of scheduling inspections and the like.

The most important employee of a general contractor is its superintendent. The superintendent is accountable for all of the trades' work on site. The very best way to foul up a project is to employ an incompetent superintendent. Knowing that, the contractor will usually be required to employ a competent superintendent and to submit the superintendent's qualifications for review in advance of starting the work. The contractor may be prohibited from employing a superintendent that the owner or designer objects to or from changing superintendents without the owner's consent.

Neither party may assign the contract without the written consent of the other party, except for assigning payments from an owner to a financial institution or other creditor. Federal or state law might restrict a contractor's ability to assign payments on some public contracts. The construction contract may carve out an exception to the assignment rule to enable the owner to assign the contract to a construction lender without needing the consent of the contractor. If not for this exception, known as a **contingent assignment**, some projects might prematurely terminate because an owner is unable to secure financing.

Performance

A contractor has to perform the work and re-perform the work if necessary to conform to the contract documents whether or not the designer has inspected, tested, rejected, accepted, or approved payment for any of the work. The contractor's warranty obligations endure long after completion of

WRITTEN CONTRACT FORM & SUBSTANCE

the work. Those obligations are fully satisfied only when the contractor's warranty expires.

Local code officials need to test, inspect and approve construction work. They will not allow contractors to do that for them. In some locations contractors are prohibited even from hiring third-party testing agencies or inspectors. Meanwhile, some building codes require contractors to pay for certain tests. Notwithstanding any such circumstances, the contractor must ordinarily arrange and pay for all test, inspections and approvals. Exceptions are drawn for any tests, inspections and approvals that were required before contract award. The owner will usually pay for those. If there are special tests, inspections or approvals that are not normally required, the contractor is usually required to arrange for these but the owner has to pay unless the special tests, inspections or approval were made necessary by the contractor's actions (or inactions). The designer has the right (but not the duty) to attend all tests and inspections and must ordinarily be given timely notice of when and where they are to occur. The contractor usually must give notice to authorities so that code inspections may be scheduled.

The reporting of errors or omissions in the design is expressed in every contract or implied if it is not. The contractor must provide prompt notice to the designer when any errors or omissions are found. The contractor could be held liable for the costs to correct errors or omissions if that cost could have been avoided by prompt notice. As it is with errors or omissions, the contractor has a duty to report any violations of applicable laws, statutes, ordinances, codes, rules and regulations that the contractor knows or should know of. Failure to report may create liability in the contractor for the cost to correct the violations.

The contractor and nobody but the contractor supervises the construction work. If the contract documents specify construction means, methods, techniques, sequences, or procedures the liability for jobsite safety remains with the contractor unless the contractor gives timely written notice to the designer about a specific safety concern and the designer instructs the contractor to proceed as specified.

Most contracts require the contractor to prepare and submit a construction schedule to the owner and designer. Construction schedules are ordinarily submitted for information, not for approval.

The contractor should keep a good, documentary record of the work at the site. Record documents include all of the contract documents plus change orders or other modifications, shop drawings, product date, samples, and all other submittals. Documents should be marked-up to show field changes made during construction. This marked-up set, known as **as-built drawings** is delivered to the designer upon completion of the work.

Shop Drawings, Product Data & Submittals

Contracts ordinarily require the contractor to prepare and submit a submittal schedule to the designer, not for information but for approval. Submittals consist of shop drawings, product data, mockups, samples and similar things. The designer reviews submittals for compliance with the contract documents. The purpose of the submittal schedule is to coordinate the handoffs between contractor and designer and to allow time for adequate design review. The designer should complete its reviews according to the approved submittal schedule. The contractor might be denied relief for delays caused by sluggish reviews if the contractor had not produced a submittal schedule for approval.

Despite the contractual conditions, contractors seldom, if ever, comply fully with submittal schedule rules. Among other reasons for noncompliance: suppliers seldom meet schedules for documentation submittal and even when they do the information they submit is frequently incorrect and has to be resubmitted.

Shop drawings are specially prepared, not simply generic, preprinted manufacturers' information. **Product data** are not specially prepared but simply generic, preprinted manufacturer's information, although they may be highlighted to indicate the model or style of product that is being provided.

Submittals do not have the same weight as drawings, specifications and other contract documents. The contract documents represent the mutual agreement between the parties. Submittals do not. Submittals merely inform the designer on how the contractor intends to implement the design. The designer may be compelled to take appropriate action with each required submittal, including instructing the contractor to correct and resubmit it. The designer, however, may not be obligated to take action on any submittals that are not required by the contract documents. The contractor should review and approve its submittals before forwarding them to the designer for approval. Subcontractors, suppliers and others are generally not authorized to send submittals directly to the designer: the contractor is usually the sole channel for communication with the designer.

Approval by the contractor is usually inferred from the act of forwarding a submittal to the designer. The contractor, however, has to perform the work (and re-perform the work if necessary) to conform to the contract documents whether or not the designer has approved submittals. Any change in the contract documents that result from the submittal process should be documented as a change in the work. The contractor should provide written notice of any new information on resubmittals so that the designer can review that information.

The contract documents may allocate specific design services to the contractor. This allocation can be expressed within performance specifications or inferred through construction means and methods. The equipment and systems that emerge from a contractor's design services must satisfy specific performance criteria. The designer, the owner, or consultants of the designer or owner develop these criteria. Contractor-provided design services are not to be performed by the contractor. They are to be performed on behalf of the contractor by a designer, engineer or other design professional that is certified and/or licensed in the jurisdiction in which the project will be built. When submittals are required for work designed by a professional who is working for the contractor, that professional must review and approve these submittals.

The contractor will ordinarily grant access to the owner and the designer, both to the site and any other locations where work is being performed, including subcontractors' shops performing sub-assemblies, remote fabrication facilities, and plants that are manufacturing customized equipment.

Protection of Persons and Property

The primary responsibility for construction safety is conferred upon the contractor. This does not relieve subcontractors of either liability for injury to their workers or their responsibility to the contractor for safety performance. Nor does this modify the owner's primary responsibility for the construction safety of other contractors working on the site. A contractor's responsibilities consist of: 1) the safety of workers; 2) safeguarding of materials and equipment, and; 3) protection of other property at and adjacent to the site. Sensible and customary safety practices are expected of the contractor: to comply with law; to provide reasonable safeguards; to designate a competent safety person; and to manage temporary loads.

The contractor <u>and</u> the owner may be strictly liable for any harm that occurs as a direct consequence of abnormally dangerous construction activities. A contractor must exercise the utmost care when performing these activities.

The contractor must fix any damage to materials, equipment, other property at the site, or property adjacent to the site, even if the damage is not covered by insurance. Adjacent property is generally covered by commercial general liability insurance. Property at the site is generally covered by builders' risk insurance. Material and equipment is generally covered by either the contractor's equipment floater insurance or the contractor's umbrella commercial general liability insurance. The contractor is responsible for damage caused by its subcontractors and anyone else who is contractually bound to the contractor. The contractor is generally not responsible for damage that is <u>entirely</u> attributable to the owner and/or the designer. The contractor may, however, remain responsible for any damage for which the contractor is partly culpable, no matter how small the fault.

In the event that potentially hazardous materials are encountered the contractor may be compelled to immediately stop work in the affected area while notifying the owner and designer in writing, unless the hazardous materials had been identified to the contractor's scope of work in the contract documents. Upon written notice of potentially hazardous materials, federal law directs the owner to retain a qualified laboratory to verify whether or not hazardous materials are actually present. If hazardous materials are confirmed the owner must remediate them (changed into non-hazardous form) or remove them. Only specially licensed contractors can lawfully perform this work. The contractor cannot.

Both the designer and the contractor ordinarily have the opportunity to object to remediation and removal contractors proposed by the owner. Work may not resume until the work area is again made safe and a written agreement to resume work is executed between owner and contractor. The owner and contractor may want to negotiate a change order for the extra costs resulting from shutdown, delay and startup. If this negotiation fails, the contractor will file a claim for damages.

The owner may be asked to add a contractual clause to compensate the contractor, the designer, and all other project participants for losses that arise from inadequate remediation or removal of hazardous materials. The owner may also be asked to compensate the contractor for any losses incurred if the contractor is held liable by a government agency for the cost of remediation through no fault of the contractor.

Because the contractor may be held liable for adverse consequences of materials and substances brought to site by the contractor, the contractor may be asked to add a contractual clause to compensate the owner from losses that arise from those consequences. The contractor should not incur liability when materials or substances that are required by the contract documents are properly handled.

In an emergency threatening the safety of people or property, the contractor must take reasonable care to protect people and property from harm. The contractor acquires this duty by virtue of being in control of the site. Expenses may be incurred and a contractor will ordinarily seek a change order or a claim for extra costs or time extensions that were necessitated by the emergency.

WRITTEN CONTRACT FORM & SUBSTANCE

The Designer

The designer must be licensed in the jurisdiction where the project is located. In most states it is unlawful for any person who is not a licensed architect to use the title "Architect."

The designer may be vested with various duties and responsibilities to provide administration of the contract. Generally, all three participants – owner, contractor, and designer – must agree to any changes to the designer's duties, responsibilities and authority. It behooves the participants to coordinate their owner-designer and owner-contractor agreements.

There are legal consequences to identifying the designer in the owner-contractor agreement: it infers the contractor's acceptance of the quality of work that the designer was known to produce. Accordingly, the contractor should have the right to object if the designer is terminated and replaced.

Administration of the Contract

The designer is ordinarily vested with limited authority to administer the owner-contractor agreement on behalf of the owner. The designer's authority to act on behalf of the owner is limited to specific duties and responsibilities that are set out in the owner-designer agreement and echoed in the owner-contractor agreement.

Typically, the designer is authorized to:

- Review and evaluate the contractor's application for payment. Inferred from this authority is the designer's right to order special testing and inspections, to reject work that does not conform to the contract documents, and to withhold or rescind certification for payment. However, authority to stop work is ordinarily vested with the owner's representative and not with the designer.
- Take actions on submittals, including instructing the contractor to make corrections and resubmit. Submittals, however, merely inform the designer on how the contractor intends to implement the design.
- Prepare change orders, construction change directives, and minor changes in the work.

- Conduct inspections for substantial completion and final completion. These inspections are separate and distinct from the designer's duty to make periodic site visits to advise the owner on work progress and conformance with the contract document.
- Interpret the contract documents and to respond to the contractor's requests for information.

The owner's representative or representatives are also authorized to act on behalf of the owner. The designer and the owner's representative(s) are different entities with different responsibilities. It behooves the designer and the owner's representative(s) to coordinate with each other.

The designer should not supervise or direct the work of the contractor or have control over construction means, methods, techniques, sequences and procedures. This proscription shields the designer from vicarious liability for wrongful acts of the contractor. It behooves the designer to avoid conduct – such as giving orders to the contractor's employees on site - that could alter this proscription. To that end, the word "administer" should not be construed as "supervise".

The designer has a professional duty to advise the owner on work progress and conformance with the contract documents. In fulfilling that duty the designer must visit the site. The designer need not be at the site full time, just visit at appropriate intervals. The designer might not find deviations from the contract document or defects and deficiencies in the work but must report what he or she observes to the owner. The contractor must inspect and test the contractor's own work whether or not the designer has inspected, tested, rejected, or accepted any work during a site visit.

Project communications are ordinarily channeled from separate contractors through the owner; from the owner or design consultants through the designer; and from subcontractors and suppliers through the contractor. The intent of theses protocols is for designer and contractor to be completely informed at all times.

Subcontractors

Subcontractors perform work on site. Working on site is the key. Material and equipment suppliers are not subcontractors unless they are doing work on site. Contractors are ordinarily required to submit, to the designer in writing, the names of all subcontractors who will perform the principal portions of the work. These names are submitted after award, although specific time limits are rarely established. The owner and designer will seek the right to object to any of them. On any objection, the contractor would ordinarily be barred from subcontracting with them. The contractor would then have to propose someone else to do the work. Contractors are rarely compelled to subcontract with anyone that they object to.

The contractor will want a change order when it costs more to subcontract with someone else. Entitlement usually hinges on whether or not the rejected subcontractor was reasonably capable of performing the work. A contractor will usually be entitled to a change order only if the rejected subcontractor was reasonably capable. "Reasonably capable" is, of course, as difficult to enforce, as it is vague.

Assignment

The owner-contractor agreement will usually require subcontractors to be bound by the contract documents in the same way that the contractor is bound to the contract documents. This imperative, known as a **flow-thru clause**, is not self-executing. Each contractor-subcontractor agreement must alert the subcontractor to the existence of a flow-thru clause. Typically, this is accomplished through **incorporation by reference** in each subcontract. Incorporation by reference makes the owner-contractor agreement an integral part of the contractor-subcontractor agreement.

Assignment is to transfer rights or duties from one party to another. A legal right confers a benefit upon the holder of that right. When a right is transferred its benefits are transferred along with the right. A legal duty exposes the holder of that duty to liabilities. When a duty is transferred its liabilities are transferred along with the duty. Contractual rights, such as the right to receive payment, can be assigned but contractual duties, such as the duty to perform work, cannot be assigned.

There is a mistaken belief that a subcontract is an assignment. A contractor cannot avoid liability by subcontracting. A contractor may delegate work to a subcontractor but the contractor still remains liable to the owner to perform the work that was delegated to that subcontractor. If, for example, a subcontractor fails to perform and then goes bankrupt, the contractor would have to complete its work and fix any defective work performed by that subcontractor, at the contractor's own expense.

There are few exceptions to this rule on assignment. One exception, called a **novation**, occurs when both parties consent in writing to transfer an entire contract, with all of its rights and duties, from one of the parties to another. Owners rarely agree to novation, however. Another exception occurs when an owner terminates a contractor for cause. When that happens, all of the terminated contractor's subcontracts are assigned to the owner, who may then reassign them to a successor contractor.

Chapter 3 PAYMENTS & PERFORMANCE

3.1 Introduction

In the common law, the duty to pay on a contract for services will not arise until the services are performed. This would occur when the contract documents are silent about payment terms. Put simply, if their contract says nothing about **progress payments**, the owner need not pay anything to the contractor until the project is completed.

Payments and Performance

Very few contractors are willing to wait until the end of a project for payment though. Owners do not usually want to wait either. The reason is that owners can obtain the lowest possible construction financing costs by offering lenders a security interest in their project. The constructor, unable to offer a security interest in the project, would have to secure financing at higher rates and pass the extra costs through to the owner. Owners who choose to make progress payments to their contractor are actually saving money.

Owners who make progress payments are also in better control. That is because payment processes provide opportunities for the owner to induce better performance from their contractor. Owners obtain leverage over their contractors by withholding payments for unsatisfactory performance. Even when a contractor is performing satisfactorily, just knowing that the owner is contractually authorized to withhold payment could goad a contractor into doing better. The owner will certainly have the contractor's attention every time a progress payment comes around.

Owners exercise indirect control over their project by choosing an appropriate pricing method. There are many different pricing methods in common use. Each is designed to incentivize contractors in different ways.

The owner's duty to disburse the contract's balance of payments, an amount that includes the profit incentive built into its pricing method, arises at the end of a project. A contractor has two hurdles to clear at the end, substantial completion and final completion. These two events are the bookends of the project closeout phase. An owner exercises control during the project closeout phase by leveraging disbursement of retention, a contractually defined portion of each progress payment that the owner withholds and accumulates to assure final completion.

3.2 Payment Processes

The contract documents will usually stipulate **progress payments**. Progress payments are periodic obligations to pay a contractor, arising as contractual conditions. Ordinarily, they are based on the percentage of work completed over a period of time, usually one month. The payment process begins when the contractor submits an **application for payment**, prepared in accordance with a **schedule of values**, to the designer. The designer reviews the application for payment and if found to be acceptable, issues a **certificate for payment** to the contractor. Finally, the contractor submits the certificate to the owner who then pays the contractor.

Schedule of Values

A **schedule of values** is a tabulation prepared by the contractor that indicates the prices for all categories of work under its construction contract. These prices include all direct and indirect costs for labor, material, and equipment, including profit and overhead. The designer refers to this schedule of values when reviewing the contractor's applications for payment. The contractor is required to submit its schedule of values to the designer before the first application for payment. The format and other requirements of the schedule of values are described in the specifications.

The architect might review a schedule of values for front loading. **Front loading** refers to increasing the prices on activities that occur early in the schedule. Prices on activities occurring later in the schedule have to be reduced so that the net amount paid to the contractor does not change. Front loading accelerates the contractor's cash flow. Early overpayment is problematic to the owner in the event of a default by the contractor.

Application for Payment

The contractor prepares an **application for payment** for work completed during the month or some other time period specified in the contract documents. The application must be submitted within 10 days or some other time period established in the contract documents, before a due date also established in the contract documents. The format and other requirements of the application for payment are also described in the contract documents.

If the parties agree to **retention** (or **retainage**), its amount and other details are specified in the contract documents. Retention is a predetermined percentage of each progress payment. Retention is withheld and accumulated by the owner as a contractual condition for assurance that the punch list will be completed and all required documents are submitted prior to final completion. Different retention percentages may be used: one for the contractor's work and another for stored material and equipment.

Subcontractors have a legal right to place a **mechanics' lien** on the owner's property in the event that the subcontractor is not paid. A mechanics' lien is a legal right to the property of the owner, created by lien statutes in the various states. Subcontractors acquire lien rights when they are not paid, even if the owner has paid the general contractor for the subcontractor's work. To reduce the risk of a mechanics' lien being placed on the owner's property, most contract documents will forbid a contractor from applying for payment for a subcontractor's work unless the subcontractor will be paid for that work.

Payments for material and equipment purchased for the work but stored offsite may be included on an application for payment but the contractor is generally required to safeguard that material and equipment and to provide evidence of the owner's clear title to it.

Certificate for Payment

The designer reviews and evaluates the contractor's application for payment. On receiving the contractor's application for payment the designer must act on it within 7 days or some other time period set out in the contract documents. The designer may choose from three actions: 1) certify the amount that has been applied for; 2) certify a smaller amount; or 3) reject the application entirely. Some contracts will authorize the designer to reconsider and reverse his or her previous certification to a smaller amount or to reject it entirely. The designer may also be authorized to adjust an item in a subsequent certification if a previous certification was mistaken.

The designer authorizes the owner to pay the contractor by issuing a **certificate for payment** to the contractor. Most contracts will include a disclaimer clause with respect to this certification. A disclaimer is a repudiation of the legal rights or claims of another. A construction contract disclaimer spells out all of the things that a designer's certificate for payment is not.

CONSTRUCTION CONTRACTS

Basically, certification is not an approval of the contractor's work. The work can still be inspected and rejected at any time.

The owner must make payment to the contractor on a payment due date set out in the contract documents provided that the contractor presents a certificate for payment to the owner. The contractor may make application for payment up to ten days before the payment due date or some other period of time set out in the contract documents. The designer has seven days, or some other period of time set out in the contract documents, to approve. Some contracts will require the payment of interest on late payments.

3.3 Pricing Methods

Owners choose pricing methods that align with their project objectives. The most common construction contract pricing methods are fixed price, cost-plus and unit-price. Both private and public construction contracts will mostly use either a fixed-price or cost-plus pricing method. Contracts for highway and infrastructure work will often use a unit-price contract pricing method.

Fixed-Price

The most common pricing method is **fixed-price**. Under a fixed-price contract the contractor receives a fixed amount of money to cover all direct costs, overhead and profit. That fixed amount is firm and not subject to adjustment attributable to the performance (or lack thereof) of the contractor during execution of the work. The Federal Acquisition Regulation [FAR] uses the phrase **firm-fixed-price** contracts to describe contracts with the federal government or agencies of the federal government that use the fixed-price method (Federal Acquisition Regulations [FAR], 1984, § 16.202). A firm-fixed-price may be modified upward for additive change order or downward for deductive change orders.

Contractors whose efficient performance results in lower costs will earn a higher profit with a fixed-price. Likewise, contractors whose inefficient performance results in higher costs will earn a lower profit with a fixed-price. There is a natural incentive for a contractor to keep his costs low in a fixed-price contract. The lower the contractor's costs the higher its profit. Conversely, the higher his costs the lower his profit. Fixed-price contracts impose no special duties on the contractor to either report costs to the owner or to suggest cost-effective and efficient procurement and construction methods to the owner

Contractors expect to bid on a fully detailed set of construction drawings and specifications - preferably prescriptive specifications - for a fixed-price. With fully detailed drawings and specifications a contractor can both accurately estimate its costs and effectively manage its fieldwork. Sometimes the contract documents will identify areas of uncertainty. Provided that a reasonable estimate of the cost of mitigating the uncertainty can be made, the owner may require a contractor to assume the risk within its fixed-price.

In certain situations, **economic price adjustment** to a firm-fixed-price contract will be allowed (FAR, § 16.203, 1984). An economic price adjustment is an upward or downward modification of a firm-fixed-price. Economic price adjustments are allowed only when certain economic events occur. So, for example, a contract for highway construction requiring much heavy, earth-moving equipment might include an economic price adjustment for diesel fuel that would give the contractor an upward or downward price modification based on changes in the average price of diesel fuel in the county where the work is being performed. This economic price adjustment would not reimburse the contractor's actual costs for fuel. Nor would it bail out a contractor who had underestimated his diesel fuel requirements. It simply provides the contractor with relief for uncertain fuel prices. The owner benefits from lower bids because bidders can avoid having to risk-load their bids to cover price uncertainties.

Economic events may include changes in published prices, changes in wage rates due to local labor agreements, and published cost indexes for prices and wages. The key characteristics of economic events are that their occurrence effects all contractors in the locality where the work is being performed and their occurrence is not within the direct control of any single contractor.

Cost-Plus

With the **cost-plus** method the owner pays the contractor its actual costs for performing the work plus a fee. The amount of this fee is determined in different ways. The amount can be a fixed sum of money, the amount can be calculated as a fixed percentage of actual costs, or the amount can be calculated as a variable percentage of actual costs. Under a **cost-plus-fixed-fee** contract, the contractor earns the same profit regardless of where his actual costs come in.

On its surface, cost-plus sounds like a good deal for a contractor. But in actuality, the amount of available profit is usually lower then it would be with fixed-price because cost-based is perceived as low-risk to the contractor. Many contractors prefer not to work cost-plus because its profits are lower. Not only that, but a contractor can even lose money under cost-plus by having to re-perform work that is found to be defective. Rework has to be done at the contractor's own expense. Put in other words: the owner is obligated to pay actual costs one time, not two times, under cost-plus.

With **cost-plus-fixed-percentage** the actual dollar amount of the fee will change with the costs, increasing with higher costs and decreasing with lower costs. If the contractor's actual costs come in high the contractor earns a larger profit. Conversely, if the contractor's actual costs come in low the contractor earns a smaller profit. Here, the best outcome for the contractor occurs when actual costs come in high. Obviously, that outcome is not the best outcome for the owner.

With **cost-plus-variable-percentage** the percentage is based on some predetermined formula. Usually, the percentage will be smaller – sometimes going to zero - for actual costs over a specified target cost and larger for actual costs under a specified target cost. This scheme takes some of the sting out of high costs for the owner but may leave the contractor without any profit. However, it does create financial incentives for the contractor to drive costs lower.

Many other types of cost-plus schemes have been devised over time. All such schemes have the common purpose of creating incentives for the contractor to reduce costs. The key point here is that the relationship between actual costs and the contractor's profit is very different under the various types of cost-plus than it is under fixed-price.

Sometimes a contractor is asked to guarantee that the price will not exceed a certain dollar amount. This scheme, a cost-plus contract with a **guaranteed maximum price [GMP]** is like a fixed-price contract in that the contractor is at risk of loss. But it is still a cost-plus contract and all cost-plus contracts, unlike fixed-price contracts, impose special duties on a contractor.

Fiduciary Duties and Cost-Plus

A **fiduciary** relationship is similar, at least superficially, to a relationship founded upon good faith and fair dealing. Unlike the relationship of good faith and fair dealing, however, a fiduciary relationship imposes a duty upon the fiduciary to serve the best interests of others, even if this results in economic or other harm to the fiduciary's interests (Starzyk, 2014).

A cost-plus contract can impose a fiduciary duty on a contractor. In a case heard in the Maryland Appellate Court it was held that an express provision in the parties' contract for the construction of a home asserted that the contractor "accepted a 'relationship of trust and confidence'" with the owners, "agreed to further their interests by performing 'the Work . . . in the most . . . economical manner consistent with' their interests," and promised "to 'keep . . . full and detailed accounts'" (*Jones v. Hiser*, 1984). The court found that the contractor had a fiduciary duty to keep the purchasers informed of the rising expenses of the house, particularly when there was no explanation of the difference between the estimated cost and the final cost. The court found that the contractor had breached this duty when he admitted that he did not know until the end of construction what the cost would be.

Cost-plus compels a contractor to suggest changes to the design that result in more cost-effective and efficient procurement and construction methods. Although these methods might reduce the contractor's profit it is legally obligated to suggest them because it is in the best interest of the owner to do so. Failure to do what is in the owner's best interest under cost-plus could very likely result in money damages being assessed against the contractor. Cost-plus tend to be more successful whenever a good business relationship had existed between the owner and the contractor prior to the award of the cost-plus contract.

Federal Acquisition Regulation & Cost-Plus

FAR § 16.306 uses the phrase "cost-plus-fixed-fee contracts" to describe contracts where the contractor will receive an amount of money to reimburse actual direct costs incurred during performance of the work plus a fixed fee for overhead and profit. This fixed fee is firm and not subject to adjustment attributable to the performance (or lack thereof) of the contractor during execution of the work. Fixed fees may be modified in response to change orders. A ceiling price will ordinarily be established that the contractor cannot exceed without prior approval from the agency. The FAR does not permit cost-plus-a-percentage-of-cost contracts (FAR, § 16.102, 1984).

Actual costs for doing work over or for uncovering work in order to correct it are disallowed in cost-plus-fixed-fee contracts. The contractor's actual costs are reimbursed only for doing the work right the first time. Lawfully reimbursable costs must be reasonable, allocable, properly accounted for, and not specifically disallowed (FAR, § 16.203, 1984). To facilitate lawful cost reimbursement, public agencies will only use the cost-plus-fixed-fee pricing with contractors who have a job cost accounting system that provides adequate detail for the agency's review. Agency personnel will routinely monitor a contractor's cost accounting and control system.

A public agency can provide **incentives** for contractors to produce better financial or non-financial outcomes than might be achieved under either fixed-price or cost-plus pricing (FAR, § 16.203, 1984).

To achieve a better financial outcome an incentive is created by first negotiating a **target cost**. The incentive (or disincentive) is applied after comparing the target cost with the contractor's actual costs at the end of the project. When the contractor's actual costs equal the target cost the contractor receives a predetermined fixed fee covering his normal overhead and profit. When the contractor's actual costs are less than the target cost, an incentive is applied to increase the contractor's fee. But when the contractor's actual costs exceed the target cost a disincentive is applied to reduce the contractor's fee. The formulas for these fee adjustments are established during contract negotiation at which time a cap would ordinarily be established for costs beyond which the contractor would proceed at his own expense.

Non-financial incentives can include early completion and performance targets such as sustainability goals, HVAC system performance or pavement strength or wear-resistance. As it is with financial incentives, fee adjustments can be applied for performance above or below a specified target, or for achieving substantial completion after or before a target date. The FAR also provides for applying multiple incentives under the same contract (FAR, § 16.402-2, 1984).

Incentive contracts are variously described as **fixed-price incentive contracts** or **cost-reimbursement incentive contracts**. For all practical purposes they are the same because they employ the same pricing methods.

Unit-Price

Unit-price contracts are common in heavy civil construction projects whenever the drawings and specification are clear and complete but the actual quantities of work to be performed are uncertain. A unit-price is a fixed price per unit of measure, for example, $14 per compacted cubic yard of soil (normally printed $14 /ccy). In this example, the contractor's fixed-price, including overhead and profit, for placing the specified soil and compacting it to specifications will be $14 for every cubic yard placed. The contractor is paid $14 times the total number of cubic yards actually placed. Ordinarily, the contractor would report quantities placed and the agency would employ a 3rd party inspector to verify quantities.

The public agency normally provides an estimate of the total quantities expected. The contractor establishes unit prices under the expectation of providing an actual quantity of work that is reasonably close to the agency's estimate. Unit-price contracts typically have hundreds of different unit-price items. A contractor's total award under such unit-price contracts is the sum of a unit-price times the quantity of work actually performed, for each and every category of work under the contract. This total award is often referred to as an **equivalent lump sum.**

3.4 Performance

Substantial performance is a contract law maxim where significant benefits of the bargain can no longer be impeded or prevented because the work of the contract has been sufficiently performed (Patrick, Beaumont, Brookie, Kirsh, Tarullo, and Spencer, 2010). Put in other words, if a material breach of contract is no longer possible, it can only mean that the work of the contract has been substantially performed. But what is a material breach?

There are two different types of contract breach: minor breach, sometimes called an immaterial breach and material breach (Eisenberg, 2002, §§816-818). The causes and consequences of a minor breach are very different from the causes and consequence of a material breach.

A **minor breach** is evidenced by failure to perform some element of the contract. An example of a minor breach by a contractor occurs when a contractor installs materials other than those that were specified. An example of a minor breach by an owner occurs when a contractor has to provide temporary electric power because the owner failed to provide it. A minor breach creates liability for damages in the breaching party. The non-breaching party must continue to perform on the contract. Typically, the non-breaching party makes a claim for damages. If the breaching party refuses to pay the claim, the non-breaching party can litigate, meaning a lawsuit to recover damages. Damages for minor breach of contract are limited to the recovery of costs.

A **material breach** is evidenced by a substantial failure to perform on a contract (Eisenberg, 2002, §§816-818). In the event of a material breach, the non-breaching party is excused from further performance on the contract. An example of material breach by a contractor occurs when a contractor abandons a project. When a contractor abandons a project the owner may stop paying, substitute a new contractor, and seek to recover from the original contractor any additional costs that arise from substituting the new contractor. An example of material breach by an owner is the owner's complete and total failure to pay the contractor. On an owner's complete and total failure to pay, the contractor may stop work, walk away from the contract, and seek to recover damages. The non-breaching party will usually litigate to recover damages. Expectation damages are recoverable for material breach of contract, consisting of actual cost, windup expenses and the expected profit that would have been earned had the project been completed.

Substantial Completion

When you perform a personal service, like painting a portrait, you must satisfy your client. Should your client not be satisfied with her portrait you would have no legal right to payment. Client dissatisfaction is tantamount to a failure to perform. Your client has no duty to pay you when you fail to perform on a personal service contract. Your dilemma is that client satisfaction is subjective and beyond your control. This is the dilemma of an artist.

Fortunately, construction work is not a personal service. It is a commercial service. Commercial services enjoy a different standard of performance (Calamari & Perillo, 1996, Ch. 5.B). Commercial services need not be satisfactorily performed; they need to be substantially performed. Substantial performance is an objectively measurable standard, not the subjective satisfaction of the owner. Nor is the performance of construction work ever perfectly true to the plans and specifications that define it (Sweet, 1997, § 9.6). Construction work need not be perfect for it to be substantially performed.

Substantial completion refers to the date upon which a contractor achieves substantial performance. Work that can be used for its intended purpose is evidence of substantial performance. A certificate of occupancy, issued by a permitting agency, is good evidence. But while a certificate of occupancy evidences usability for intended purpose it is not interchangeable with substantial completion. The criteria for an occupancy permit vary with local authorities while the criteria for establishing substantial completion is fixed by the contract documents and triggered by substantial performance.

When the contractor notifies the designer that the work is substantially performed it starts a chain of events that lead to substantial completion. Upon notification the designer will then schedule inspections to ascertain whether the work is substantially performed or not. If, upon completing the inspections, the work is found to not be substantially performed yet, the contractor must correct the work indicated by the designer and request another inspection. This process can repeat over and over.

Substantially performed work is not perfectly performed work. Work need only be performed to an acceptable and objectively measurable standard in order to be substantially performed. And work that is substantially performed is far from being finished. There are always many things remaining to be done. None of this remaining work, if left undone, would constitute a material breach. But all of the remaining work still needs to be finished.

Substantial performance is usually accompanied by a **punch list**. A punch list is a tabulation of all work items that still remain to be finished after substantial completion. The contractor and designer prepare the punch list during a walk-thru at the time of the substantial completion inspections.

If the designer finds that the work is substantially performed, he or she issues a **certificate of substantial completion**. The certificate establishes the substantial completion date along with the responsibilities of the owner and contractor for finishing the items on the punch list. The punch list should be attached and both owner and contractor should sign the certificate. The owner may be authorized to release withheld retainage at substantial completion, up to an amount needed for incomplete work, deficient work needing correction, or damages for delay in achieving substantial completion, if any.

Conditions might be established in the contract documents under which the owner may occupy or use part of the work before it is substantially complete. The owner and contractor will ordinarily sign a separate agreement for early occupancy, which would include provisions for damage to work finished or unfinished as a consequence of early move in. The property insurer must approve any agreements for early move in.

Final Completion

The substantial completion inspection is one of only two mandatory inspections by the designer. The other is the **final completion** inspection. In the final completion inspection, the owner and designer walk thru the project to determine that all items on the punch list are finished.

When the contractor has finished the work on the punch list, demobilized the site, and all documents and supporting documents are prepared, the contractor submits to the designer:

- A final application for payment;
- All required documents such as instruments of service, as-built drawings, manufacturers warranties, operating manuals and spare parts lists;
- Supporting documents; and
- A request for final inspection.

CONSTRUCTION CONTRACTS

The term **instruments of service** refer to studies, surveys, models, sketches, drawings, specifications, and other things created in writing, in physical form, or in electronic representations by the designer. It is a broadly defined term that is intended to capture every means of embodying the services that a designer provides. Instruments of service are not automatically contract documents. Some are. Some aren't. But they are all protected by copyright. The contractor is licensed to use instruments of service for the purpose of constructing the work but once the work of the contract is complete the instruments of service must be returned to the designer.

Supporting documents may include:

- An affidavit that all debts connected to the work have been paid (An affidavit is a declaration of facts written down and sworn to in front of a notary public or other legally authorized person);
- Certificates of insurance showing that all coverage is currently in effect; and that required coverage will remain in force after final payment and will not expire or be canceled without giving prior written notice to the owner;
- A written statement that the contractor knows of no substantial reason that insurance will not be renewable;
- Consent of the surety (if surety bonds are in place) to final payment; and
- Other evidence of payment that is required by the owner.

Upon receiving the contractor's submittal, verifying the receipt of all documents, confirming the amount of final payment, and validating all of the supporting documents the owner and designer will then conduct a final completion inspection. If all of the work on the punch list is finished, the designer will then issue a **final certificate for payment**. The contractor receives final payment from the owner a specified number of days after the owner receives his final certificate for payment.

The owner waives most rights to recover damages from the contractor – there are exceptions for unsettled claims - by making final payment to the contractor. There are exceptions that preserve the rights of the owner insofar as latent acts, omissions, or defects, warranty rights, and other rights that survive final completion. If there are any unsettled claims at the time of final completion, the contractor is expected to restate them specifically, in writing, at that time.

Substantial Completion and Liquidated Damages

The distinction between substantial completion and final completion is especially important if there are **liquidated damages** for late completion. Liquidated damages are a good faith estimate of actual damages for late completion, usually expressed as a certain amount of money per day, or *per diem,* Latin words meaning "each day". Liquidated damages are assessed against late substantial completion, not against late final completion. The substantial completion date is an important date because a contractor can no longer be terminated for cause (termination for cause is a material breach of contract) once substantial completion is achieved; the owner's warranties start to toll on the substantial completion date; and the substantial completion date is the date for which liquidated damages are calculated.

Chapter 4   CONCEALED OR UNKNOWN CONDITIONS

4.1 Introduction

Concealed and Unknown Conditions

Subsurface or other conditions on site may prove to be other than those that were expected. Archaeological artifacts may be uncovered. Unexpected environmental hazards may appear. Who bears the risk? To answer that question the parties look to their contract. An important function of every construction contract is to allocate risk. Constructors deal with risk by evaluating the risks that a projects presents; looking to their contract to find out who is accountable for that risk; and deciding whether to hold, assign, reduce, transfer, or share each risk that has been allocated to them.

This chapter covers risk allocation for concealed or unknown conditions.

4.2 Historical Background

Concealed or unknown conditions, known also as **differing site conditions**, **differing conditions**, or **changed conditions**, occur when the conditions that a contractor actually finds on a construction site during construction are not the same as what the contractor expected to find at the time when the contract for construction was formed (Kelleher, 2005, p. 211). Subsurface soil conditions are the most common type of concealed or unknown conditions but there are many other types. Asbestos containing materials, lead-based paint, surprisingly thick concrete, unanticipated steel reinforcement in buildings that must be demolished, and unsuitable material in borrow-pits are all concealed or unknown conditions.

Concealed or unknown conditions are risks. Our first question is then: when a contract is silent about concealed or unknown conditions, who bears the risk?

Paradine v. Jane

A contract creates a duty to perform the work of the contract and it has long been established in common law that the contracting party who performs the work of the contract assumes all risks to performance. The legal precedent for this was set long ago, in *Paradine v. Jane* (1642).

Paradine emerged out of the English Civil War during the reign of King Charles I. Armies on both sides of that conflict frequently looted the land in order to gain supplies. As it happened, Paradine had leased land to Jane who was expelled from that land by an invading army. Jane was unable to earn any income from the land during the army's occupation. In three years, however, armed resistance collapsed, the land was relinquished and Jane took back possession. Paradine then demanded payment of back rent that Jane refused to pay. Paradine sued Jane and was awarded back rent by the court, who reasoned that Jane had gotten what he had contracted for - possession of the land – and for getting what he had contracted for Jane had a duty to pay. Jane's duty to pay did not go away just because somebody else took possession away from him.

Paradine may appear to be a harsh judgment. Yet the court's harsh judgment was reasonable and necessary. What mostly concerned the court was that all of the good that comes to the public from enforcing contracts could be jeopardized were the court to start chipping away at the basic premise: when you enter into a contract you assume all of its risks. Although U.S. common law has made a few encroachments into this rule, *Paradine* survives to this day. You can count on it being applied, with very few exceptions (Sweet, 1997, §8.6).

Liability

One of the contracting parties will be liable for concealed or unknown conditions. **Liability** is a legal obligation or accountability that must be satisfied at the expense of the liable party. The basic question is this: who will be liable for the extra costs of performance incurred as a consequence of finding concealed or unknown conditions? When the contract documents are silent on that question, the courts will follow the *Paradine* rule: the contractor assumes all of the risks of performance. In other words, the contractor has to pay, no matter the cost. A finding of contractor liability for concealed or unknown conditions is never a good outcome for the contractor. The extra costs of concealed or unknown conditions can potentially ruin a contractor.

A different spin is put on the basic question of liability for concealed or unknown conditions when relevant information – geotechnical report, plans and specifications describing an existing building, or a recommended borrow pit – is given to the contractor prior to bidding. Will the contractor's reliance on that information excuse the contractor from liability if that information proves to be mistaken?

The *Spearin* Doctrine

The ***Spearin* doctrine** emerged from the case of *U.S. v. Spearin* (1918). Because the U.S. Supreme Court decided *Spearin*, it applies in every state. *Spearin* created an implied warranty that plans and specifications provided by the owner are free of design defects known to the designer. This says that if the designer knew of particular design conditions (for example: soil type, water table, weather conditions, electrical power characteristics, unusual site drainage), failed to incorporate those conditions into the design and the contractor was thereby misled, the owner is responsible for the consequences. The designer is charged with knowing what a competent designer is expected to know. As long as the designer could have reasonably been expected to know something he is charged with knowing it whether he actually knows it or not.

The *Spearin* doctrine changed everything. It shifted liability for concealed or unknown conditions from designers to owners After *Spearin*, contractors could simply pursue a warranty claim against the owner. To succeed at that claim it is only necessary to show that misinformation was provided; that the misinformation could not have been discovered by a competent contractor in a site visit prior to bidding or that the contractor was denied the opportunity of a site visit prior to bidding; and that the contractor relied upon that information in preparing his bid (*U.S. v. Spearin*, 1918).

Since *Spearin*, courts have also held that contractors are not liable for concealed or unknown conditions where it could be shown that the owner had relevant information that it should have disclosed but did not disclose to the contractor.

CONSTRUCTION CONTRACTS

4.3 Modern Rules

The FAR provides a model for concealed or unknown conditions that has gained widespread acceptance in both the public and private sectors. Employing the term **differing site conditions**, FAR §52.236-2 establishes rules for differing site conditions, as follows:

(a) The Contractor shall promptly, and before the conditions are disturbed, give a written notice to the Contracting Officer of:

(1) Subsurface or latent physical conditions at the site which differ materially from those indicated in the contract; or

(2) Unknown physical conditions at the site, of an unusual nature, which differ materially from those ordinarily encountered and generally recognized as inhering in work of the character provided for in the contract.

(b) The Contracting Officer shall investigate the site conditions promptly after receiving the notice. If the conditions do materially so differ and cause an increase or decrease in the Contractor's cost of, or the time required for, performing any part of the work under this contract, whether or not changed as a result of the conditions, an equitable adjustment shall be made under this clause and the contract modified in writing accordingly.

(c) No request by the Contractor for an equitable adjustment to the contract under this clause shall be allowed, unless the Contractor has given the written notice required; provided, that the time prescribed in paragraph (a) of this clause for giving written notice may be extended by the Contracting Officer.

(d) No request by the Contractor for an equitable adjustment to the contract for differing site conditions shall be allowed if made after final payment under this contract (FAR, 1984).

FAR §52.236-2(a) requires the contractor to supply prompt notice but stops short of setting a time limit (FAR, 1984). In any event, once discovered, the conditions should not be disturbed. Most private construction contract documents, such as the standard form construction contracts published by the AIA, will require the contractor to provide notice within some period of time, typically 21 days, from first observing differing conditions. Although the AIA does not specify the form of notice, a prudent contractor should evidence its notification in writing.

FAR §§52.236-2(a)(1) and (2) provide what has come to be the most widely circulated definition for differing conditions, establishing two separate types, commonly known as type I and type II differing conditions (FAR, 1984). Type I differing conditions arise from mistaken information communicated by the contract documents, while type II differing conditions flow from common knowledge tacitly held within the construction industry.

Type I Differing Conditions

A type I differing condition arises when what is found during performance of the work is at variance with information that was either expressed in the contract documents or that was reasonably inferable from the contract documents. To be compensated for a type I differing condition a contractor must show that:

1. Certain conditions were expressed in or could be inferred from the contract documents and the contractor relied on those conditions;
2. Actual conditions found while performing the work were substantially different than the conditions that had been relied upon;
3. Timely notice of differing conditions was given; and
4. The contractor suffered delays and/or extra costs as a direct consequence of those differing conditions.

A contractor's ability to be compensated for type I differing conditions hinges upon being able to identify the mistaken information or inferred information in the contract documents and to establish that the contractor could not have known that that information was mistaken by any reasonable means at the time that the contract was formed.

Type II Differing Conditions

A type II differing condition arises when what is found during performance of the work is unusual or at variance with what is normally encountered. To be compensated for a type II differing condition a contractor must show that:

1. Certain conditions were reasonably anticipated for the work of the contract in the location where the work is being performed;
2. Actual conditions found while performing the work were unusual and substantially different that what was reasonably anticipated;
3. Timely notice of differing conditions was given; and
4. The contractor suffered delays and/or extra costs as a direct consequence of those differing conditions.

A contractor's ability to be compensated for type II differing conditions hinges upon establishing what could reasonably be anticipated from the contractor's experience working at the site of the project, or from experience doing similar work, or from common knowledge within the construction industry at the time that the contract was formed.

FAR §52.236-2(b) requires a prompt investigation and recommendation of an equitable adjustment in price and time; either up or down as the case may be (FAR, 1984). The word equitable is used in the common law for *"just and consistent with principles of justice and right"* (Garner, 2006). Basically, a fair price and time extension should be recommended. The contractor should file a claim if the recommended price and time are not equitable.

Allocation of Liability

Contractors are expected to be diligent and competent in reviewing information provided to them by the owner. Because of that, liability for concealed or unknown conditions will remain with the contractor who, having had the opportunity of a site visit prior to bidding, either failed to identify obvious misinformation or identified obvious misinformation but failed to notify the owner. Liability will also remain with the contractor when the contract documents expressly say that the contractor will be liable for differing conditions.

In summary, either the owner or the contractor assumes liability for concealed or unknown conditions through the express terms of the contract between them. However, whenever the contract is silent about concealed or unknown conditions, liability remains with the contractor unless the owner provides relevant information. In such circumstances, the following rules apply:

The **contractor** will be liable for concealed or unknown conditions when:
- The relevant information is accurate;
- The contractor could have identified misinformation during a site visit but did not do so, having had the opportunity of a site visit; or
- The contractor had discovered the misinformation in a site visit but failed to notify the owner about the misinformation prior to bidding.

The **owner** will be liable for concealed or unknown conditions, when:
- The contractor could not have identified misinformation during a site visit and the contractor relied on that misinformation in forming his bid; or
- The contractor was denied the opportunity of a site visit and the contractor relied on misinformation when forming his bid; or
- Information in the possession of the owner was NOT revealed to the contractor but had it been revealed the contractor would have changed his bid.

Disclaimers

Even with those rules, however, an owner can avoid liability for misinformation by expressing a disclaimer in the contract. Such disclaimers generally say that the owner is providing information to the contractor only as a courtesy and does not take any responsibility for the accuracy of that information. Courts will generally enforce such disclaimers (Sweet & Schneier, 2013, §20.04).

Burial Grounds, Archaeological Sites & Wetlands

Burial grounds, archaeological sites, and wetlands that are not contemplated in the contract documents but discovered during construction would be considered ordinary concealed or unknown conditions except that they have to be treated differently because federal and state regulatory agencies become involved. Typically, a statutory duty to take action is placed on the owner. Upon discovery the contractor should immediately suspend construction operations and notify the designer. Although federal law does not specify the form of notice, the prudent contractor should evidence this notice in writing. Construction operations cannot resume until the owner obtains governmental authorization to resume.

Under these circumstances, the contractor would ordinarily file a claim for adjustment of price and schedule.

Chapter 5   CHANGES

5.1 Introduction

The common law will enforce a contract as it was originally written unless both parties agree to change it (Kelleher, 2005, p. 186). This could pose a dilemma for an owner who may want to change the scope of work and is willing to pay for it but the contractor simply refuses to do it. It could also happen that the owner wants something removed from the scope of work (a deductive change) but the contractor insists on being paid the originally contracted for amount. Under either circumstance, the owner would have no recourse because the common law would enforce the original agreement.

If both owner and contractor want to change their contract, they can do that but their change must satisfy the preexisting duty rule. To satisfy that rule, sufficient new consideration must accompany the change. For example, an owner cannot order a contractor to do more without paying the contractor more (or providing some other benefit to the contractor). Nor can a contractor demand more money from the owner without providing more service to the owner (or providing some other benefit to the owner). Sufficient consideration must accompany the change. As long as both parties agree to a bargain that has sufficient consideration, the change to their contract will be enforced.

5.2 The Changes Clause

A good construction contract will anticipate the change dilemma – where one party wants to change the contract but the other does not - and it will solve it with a **changes clause**. A typical changes clause empowers an owner to order a change in the work under which the contractor is instructed to proceed with the change and produce, in good faith, an estimate of costs for owner approval. Under the typical changes clause, the owner cannot withhold his approval without good cause and must pay the contractor for all approved changes.

Sometimes there is a dispute over the cost, which frequently results in a claim. A claim is simply the name that contractors use for a circumstance where a contractual dispute becomes a demand for money. But claims issues aside, the contractor simply cannot refuse to change the work when there is a changes clause in his contract.

Now suppose that a contract has been written but one of the parties fails to or refuses to sign it. If the work has started the court may determine that there is an enforceable contract: the partially signed writing providing evidence of the substance of the bargain. But only the essential substance of the bargain – in other words the specific thing that is being built and its price – will survive. All other terms in the writing are usually discarded, including any changes clause. What remains is a contract that, in the absence of any written clauses, defaults to common law rules. Among those rules, the court will not enforce any change to the original contract unless, of course, both parties agree to a change and sufficient consideration exists to support that change. So when one party fails or refuses to sign the written contract, the basic bargain survives but neither party has the power to change the basic bargain without the agreement of the other.

The standard form construction contracts published by the AIA, the EJCDC, and ConsensusDocs™ all contain changes clauses.

Cardinal Change

The courts have found limitations on what a contractor can be compelled to do under a changes clause. An owner cannot order an extreme change to the scope of work. Such an extreme change, known as a cardinal change, can be defined as a major undertaking, substantially different from anything originally contemplated in the contract (*County of Cook v. Henry Harms*). If, for example, a contractor had a contract to install sidewalks in a subdivision, a changes clause could not be used to order that contractor to put in the roadways and the foundations for all of the houses too. That sort of substantially different and major undertaking would be a cardinal change. But if, for example, the owner wanted to increase the width of all of the sidewalks from 4 ft. to 6 ft., that would be an acceptable change under a changes clause.

Contractor-Subcontractor Agreements

Suppose that a general contractor and an owner have executed a good, written agreement, and that agreement includes a changes clause. During the course of the work the owner orders a change that removes some of the work of an electrical subcontractor. The general contractor can find himself in a difficult spot if the electrical subcontractor had started work without signing his subcontract agreement. Because while the general contractor has no choice but to instruct the electrical subcontractor not to perform that part of the work – the general contractor would be compelled to do that by the changes clause in his agreement - the electrical subcontractor cannot be compelled to reduce his price without his consent to the change. Without the electrical subcontractor's consent to a change, the court will enforce the electrical subcontractor's original basic bargain. The practical consequence of this is that the general contractor will have to pay his electrical subcontractor the originally contracted for price regardless of how much work is removed from his scope.

Practical Advice

Look for a changes clause in your written contract. If there isn't one and you are still negotiating the contract, try to get one inserted. Be sure to add instructions for how to handle deductive changes. Deductive changes are changes that remove items from the scope of work or otherwise reduce the cost of the work. Many a claim involving deductive changes hinges on indirect field costs, general overhead costs, and profit. Should you give some portion of each of those back to the other party on a deductive change? Or not? It is better to decide before work starts. Whatever it is that you and the other party agree to, insert it into your contract. Put changes clauses in your subcontracts. Make sure that all of your contracts are signed, particularly your subcontracts, before work begins.

5.3 Contractual Processes

The processes of the AIA standard form construction contracts provide choices: a **change order** if the owner and contractor agree on the amount of extra money and time required to effect a change in the work; a **construction change directive** when the amount of extra money and time has not yet been determined; and a **minor change** in the work where neither extra money nor extra time is involved. Every change must be accomplished as a minor change or change order. **Substitutions** may also be framed as changes. A substitution is the use of a different material, method or technique than what was specified, usually by request of the contractor with approval of the designer. A clause is usually inserted that prevents a contract from being changed by the action or failure to act of either of the parties. In the common law, if a contract is in writing, changes to the contract must also be in writing.

Change Orders

Change orders describe modifications to the work of the contract, states the change in the contract amount and the schedule extension, if any, necessitated by the modifications. If work is removed from the scope of supply the contract amount is lowered, the schedule is reduced, if appropriate, and the change is called a **deductive change**. Among other things, change orders have to be in writing and the owner, designer and contractor must sign them. To avoid misunderstandings, the words "no net change" should be inserted if either price or time or both are unchanged.

Construction Change Directives

A **construction change directive**, or **interim directed change**, gives the owner the right to order changes to the work. Both the owner and the architect sign these orders. The contractor does not. The directive is effective on receipt without the contractor needing to sign. The owner may order a change even when the contractor and designer and/or the owner disagree on price and other terms of the change. A directive is also used when there has been insufficient time for the contractor to develop a price.

The owner's rights are not unlimited: the common law bars cardinal changes. Directives constituting major modifications to the work should first be submitted to the surety for approval because the owner does not have the right to order the surety to modify payment or performance bonds. It is also important to note that both parties must proceed in good faith that a mutually acceptable price and schedule can be found. By signing a directive the owner becomes obligated to pay for the work.

There are many methods for pricing change directives. These methods may also be used on change orders. If unit prices are stated in the contract documents they may be applied to change directives. The contractor is usually obligated to submit accounting records and supporting data to the designer. When a suitable price cannot be arrived at, the designer may develop a price by calculating all costs and adding profit and overhead.

The designer may make an interim determination whenever the contractor has completed some or all of the work under a directive, but the price has not yet been settled. The amount of that interim determination can be certified for progress payment as a temporary adjustment until the price of the directive is settled.

Disputes on equipment rental rates can be avoided by stating their basis in the owner-contractor agreement. The contractor may be required to give back only actual costs, not profit and overhead, for deductive changes. And at the end of the day, the parties must produce a change order when the amount of a directive is finally agreed to.

Minor Changes in the Work

The designer may issue a unilateral written order for a **minor change**, or incidental change, in the work. Minor changes do not involve adjustment in the contract price or time. The designer should not order minor changes that are inconsistent with the contract documents. Inconsistencies should be handled as change orders or directives adding the phrase "no net change" to price and time. To avoid disputes, the designer should also seek the contractor's agreement that price and time are not affected by the minor change being proposed.

5.4 Preexisting Duty Rule

Suppose a licensed contractor is remodeling your kitchen. You have a written contract for a fixed price of $30,000. Your contract complies with all of the statutory laws in your state. It is a good, entirely legal, enforceable contract. Half way through the job, your old kitchen demolished, your contractor demands $10,000 more to complete the job. He just demands $10,000 more and says that if you don't sign his change order he is walking off the job. You object, of course. But fearing that if he walks off the job your kitchen will never get put back together, you begrudgingly sign his change order. You don't like it. But what can you do? The price was supposed to be $30,000, no more. Now that you have signed a change order you have to pay $40,000. Or do you?

Under the circumstances that have been described, this change order is not enforceable. The reason for that is that common law requires sufficient new consideration to accompany any change to a common law contract. Let's examine this change order to see if it provides sufficient new consideration. Here are the facts. Your contractor gets $10,000 more but gives up nothing in return. You pay $10,000 more to get nothing new in return. For there to be sufficient consideration both parties must get something of benefit and give up something as detriment. That does not happen given the facts as they are described here. So this change order does not provide sufficient new consideration.

This example illustrates the **preexisting duty rule**. This rule says that no contract can be formed in circumstances where a person's duties after a promise are exactly the same as they were before the promise (Eisenberg, 2002, §§ 6-7)(*Ligenfelder v. Wainwright*, 1891). In our example, the contractor had a preexisting duty to remodel the kitchen for $30,000. Unless he does something different – upgrade the floor tile, add new cabinets, something – he cannot get any more money. It does not matter that both parties signed the change order. It is unenforceable for lack of new consideration. So ignore it and do not pay that additional $10,000.

It is also instructive to know that judges can tell the difference between real consideration and fake consideration. Were you to try to get around the Preexisting Duty Rule by offering to do something different that had little or no real value, you would not be successful. A pretense of consideration, known as "nominal" consideration is insufficient consideration. A good example of nominal consideration is a contract to do something for one dollar. Nominal consideration is insufficient to form an enforceable contract and it is insufficient to form an enforceable change to a contract.

The Uniform Commercial Code

The **Uniform Commercial Code [UCC]** applies to contracts for the purchase of goods (material and equipment), whereas the common law applies to contracts for the purchase of services (labor) (Kelleher, 2005, p. 103). Changes are enforceable <u>without</u> new consideration under the UCC (White & Summers, 2000)(Amer. Law Inst., 2011). Our Preexisting Duty example would be invalid if the court found that our contract was a UCC contract and not a common law contract. Would the court find our contract to be UCC contract if our job required mostly goods and few services? The answer is no. Despite the amount of material it takes to remodel a kitchen, the common law considers those materials to be incidental to providing the service of remodeling. The courts almost always classify construction as a service, regardless of how many goods are involved; therefore, construction contracts are almost always governed by the common law and not the UCC (Eisenberg, 200, §§ 6-7). You should always depend on needing new consideration as a requirement of enforcing any change order.

Chapter 6   TIME

6.1   Introduction

The typical construction contract exchanges a promise to build for the promise to pay. That is the basic bargain. But what does that basic bargain say about time? Is time a part of this basic bargain? This chapter explores questions of time: whether or not time is a performance obligation in a construction contract and if it is, how to allocate liabilities as a consequence of not performing on time.

6.2 Performance

People must agree that time is a performance obligation if failure to complete on time constitutes a material breach from which damages for late completion will flow. The words that express this agreement are: **time is of the essence.** Time may not be a performance obligation unless your contract says that time is of the essence. Unless time is of the essence, the date of completion will be construed to be "a reasonable time". It could then become necessary for a jury to make a factual determination of whether a project's actual completion date was reasonable or not.

A contractor is typically given notice to proceed with construction on a date certain with the expectation that the work will be completed by a certain future date. If the contractor does not complete the work by the certain future date, the contractor's performance will be late. If time is of the essence then the contractor will be liable for the consequences of late performance.

This raises the question: if time is of the essence will the contractor be liable for the consequences of early performance? An owner would certainly be happy with any contractor who completed a little early. But what would happen if the contractor completed the work long before its completion date? Let's suppose that on a one-year project the contractor was to com-

CONSTRUCTION CONTRACTS

plete it five months early. Will the owner have money to pay the contractor when the progress payments come due five months earlier than planned? Early completion can create a serious cash flow problem for an owner.

Now let's think not about an early finish but an early start. Suppose that the contractor started the project early. In that event we have the same payment problem if the early start results in an early completion. We also have a new problem. The contractor's early start could ruin the owner's chances of getting construction financing. The reason has to do with security interests.

Security Interests

Most owners will secure a construction loan to cover progress payments during construction. A construction lender expects that an owner will pay off their construction loan with a mortgage loan that the owner will obtain after the work is substantially complete. A mortgage is a legal device that gives a lender a legal right of title to property as security for repayment of a loan. A construction lender relies on an owner to obtain a mortgage loan. But just in case something goes wrong and the owner cannot obtain a mortgage loan, the construction lender wants to be able to force a sale of the building to recoup their loss. The legal process of forcing the sale of a building to recoup a loss is known as foreclosure. When a lender has the contractual right to foreclose on a building, the lender is said to have a security interest in that building. Security interests are created by contractual agreement between the parties, usually through the written terms of a construction loan agreement or a mortgage.

Contractors also have a security interest in what they are building. They get that right through a statutory law device known as a mechanics' lien. Basically, a mechanics' lien gives a contractor the right to foreclose on a building to secure payment if the contractor has not been paid. A lien is like a security interest except that mechanics' lien processes and other operations of law will create a lien, not a contractual agreement between the parties. This is a subtle difference, to be sure. So in practice, the terms security interest and lien are used interchangeably.

Every state has their own mechanics' lien statutes. Generally, these statutes provide both rules that create lien rights in the contractor and complex legal processes to secure payment. The rules and processes differ from state to state but the basic purpose of all mechanics' lien statutes are the same: to secure payment for a contractor who has not been paid.

Let's get back to our construction loan question. What happens when the construction lender and a contractor <u>both</u> have a security interest against the same building? There is only one building to sell in foreclosure and usually not enough money from that sale to pay off everybody standing in line to get paid. By law, the first person in time with a security interest against a building is the first person in line to get paid. The first person in time is the one who started first. So when a contractor jumps the gun on the notice to proceed and starts work before the construction lending is secured, the contractor becomes the first person in time (Sweet, 1997, §9.3). Potential lenders, seeing that happen, might withdraw from lending negotiations with the owner. The consequence of that could be an owner's inability to pay for the building's construction and from that a breach of the owner's promise to pay the contractor.

So it is easy to see that an owner might suffer financial damage when a contractor either starts early or finishes especially early. But does the typical contract prohibit the contractor from starting early or finishing early? Earliness is certainly not prohibited in a verbal contract, an inchoate contract or any contract that is silent about time. The fact is that not even the standard form construction contracts expressly prohibit early completion by a contractor. A contractor who completes especially early may still be compelled by obligations of good faith and fair dealing to notify an owner of his intent to complete early (Sweet, 1997, §9.4). And owners could be liable for damages for preventing a contractor from completing early, whenever the contractor had notified the owner of his intent (Sweet, 1997, §9.4).

Late Performance

What about the more common scenario where the contractor is late? Generally the contractor would be excused for late completion if it can be shown that the owner caused the delay. The contractor, of course, is unexcused if the delay is attributable to the contractor. But even if late completion were attributable to the contractor, to be awarded damages the owner would still have to prove two things: 1) that the owner suffered a certain dollar amount of money damages; and 2) that certain dollar amount of money damages was the direct consequence of the contractor's unexcused, late completion.

Late completion will certainly cause loss of use of a building. But it is often difficult to convert loss of use into a certain dollar amount of damages, especially in residential construction. What if the opposite thing happens: the owner is late in paying the contractor? The contractor would also have to convert late payment into a certain dollar amount of damages. This is a very difficult thing for most contractors to do. Late completion by either party is meaningless as a breach of contract unless damages measurable as a certain dollar amount of money can be proved to be the direct consequence of that breach. The bottom line is that it is not easy to show actual damages when either of the parties to a contract performs late.

Liquidated Damages

Even when the contract states that time is of the essence, the obligation will still be on the aggrieved party to prove actual damages when the other party is late in performing (Sweet & Schneier, 2009, §26.09A). To avoid the difficult problem of proving actual damages, some owners will insert a liquidated damage clause into their contract with the contractor. Liquidated damages are a good faith estimate of actual damages for late completion, usually expressed as a certain amount of money for each day of unexcused late completion (Kelleher, 2005, §16.26). The Latin words that express an amount of money for "each day" late are **per diem**. As long as the estimated *per diem* is reasonable; was based upon a good faith estimate by the owner of actual damages; and was included in the contract at the time that it was agreed to, the court will enforce it (Kelleher, 2005, §11.42).

The courts will not enforce a penalty. A penalty is an unreasonably large amount of money, assessed against a contractor to intimidate the contractor into timely performance. Penalties are not enforceable. Yet if a penalty is called out in the liquidated damages clause of your contract, the court will not ignore the penalty. The court will first try to determine if the word "penalty" in your contract has the same meaning as the word "penalty" in the common law. The key issue is the *per diem* amount. Is it a reasonable estimate of what the owner stands to actually lose due to late performance? If it is, then the *per diem* becomes the owner's compensation for the contractor's late completion. But if the *per diem* is not a reasonable estimate, if it is just there to intimidate the contractor into timely performance, then the *per diem* amounts to a penalty and the court will not enforce it.

Courts will enforce liquidated damages at the *per diem* stipulated in the construction contract, even if it can be shown that actual damages were either higher or lower (*Georgia v. Norair*, 1973). The courts of some states will, however, construe liquidated damages that are excessive or not based upon a good faith estimate as unenforceable penalties (*Southeastern v. Real Estate World*, 1976).

California statutes - presuming that liquidated damages clauses are valid - shift the burden to defendants to prove that liquidated damages are excessive or not based upon a good faith estimate (*Jurkovich & Hebesha*, 2009, p. 157). Courts in other states think similarly (*Farmers v. M/V Georgis Prois*, 1986)(*Hubbard v. Lincoln*, 1986)(*Coe v. Thermasol*, 1985-

1986). When all is said and done, it is difficult to prove that liquidated damages are excessive or not based upon a good faith estimate. Therefore, liquidated damages generally hold up in California courts.

Liquidated damages clauses are unpopular with contractors. So they try to avoid them. But before a contractor negotiates the liquidated damages clause out of his contract, he should carefully consider the risk that he is taking. When time is of the essence, a contractor will be liable for actual damages that are the consequence of his unexcused late completion. This liability does not go away just because a liquidated damages clause is absent from his contract. And although it is difficult for an owner to prove actual damages, it is not impossible to prove them. The problem for the contractor is that actual damages might turn out to be much higher than the liquidated damages would have been. Liquidated damages work to a contractor's benefit because they create a cap on his liabilities. The courts will usually enforce the *per diem* even if the actual damages are much higher. A contractor takes a risk with a contract that does not have a liquidated damages clause. He risks paying actual damages. When you think about it, it's a wonder that more contractors do not insist on having a liquidated damages clause in their contract.

The Anomalies of Time

When a construction contract says time is of the essence it becomes a performance obligation. Yet the consequences of breaching that obligation are often unclear. Late completion gets a lot of attention, much of which is focused upon liquidated damages. Liquidated damages generally hold up in court but not if they are construed as a penalty. And when they do hold up they act as a cap on actual damages that, ironically, work to the benefit of the contractor. Contract language may prohibit – to prevent unwanted security interests - an early start by the contractor. Early completion, however, rarely gets much attention even though it can create serious cash flow problems for an owner. A contractor should at least notify an owner of his intent to complete early. Finally, an owner should have contract provisions that bar early completion in order to avoid liability for damages for preventing a contractor from completing early where the contractor had notified the owner of his intent.

6.3 Liabilities

By signing an owner-contractor agreement a contractor accepts the time allowed for construction in the contract documents and cannot later claim that that time was insufficient. The contract documents will stipulate that the **date of commencement** of the work is the date of the owner-contractor agreement, some other date, or a date announced in a notification to proceed that will be sent to the contractor on a future date. They also stipulate that the expected **substantial completion** date is either a date certain or a certain number of days from the date of commencement. Time will start to toll on the date of commencement, whether the contractor has actually started the work or not. Most contracts will assert that the work must be substantially performed by the substantial completion date or a later date should any changes to the contract occur that justify a time extension.

Delay

Excusable Delay

An **excusable delay** is an impediment to the work that arises from certain contractually listed causes. Contractually listed causes for excusable delay can be organized into five broad categories, as follows:

1. An act or omission by the owner or designer, including employees of either and separate contractors employed by the owner;
2. Change orders and construction change directives;
3. *Force majeure* events;
4. Delay authorized by owner pending mediation and arbitration; and
5. Other causes that the designer determines may justify delay.

Every excusable delay must be coupled to a project's critical path. A project's **critical path** is a subset of project activities - each activity in the subset is a **critical activity** - arranged in a chronological sequence that establishes the shortest time path from the date of commencement to the substantial completion date. For a delay to be excusable it must hinder a critical activity as a direct consequence of one or more of the justifiable causes that are specified in the contract documents.

Continuous delays are excusable delays that are interrupted from time to time although recurring for the same underlying cause. Both excusable delays and continuous delays will give the contractor more time to complete the work but they will not usually justify an increase in price. When there is no increase in price associated with a continuous or excusable delay, they are classified as **noncompensable delays**.

The contractor is generally required to initiate contractual processes by making an excusable delay claim. The designer then evaluates the claim and decides if an extension is justified by the contract documents and the circumstances of the delay. If the designer determines that the delay is justified he determines its extent and prepares an appropriate change order.

Force Majeure

The words *force majeure* literally mean "higher force." It refers to acts of God, acts of the government or the military, labor disputes, transportation disruptions, infrastructure failures, and other things disrupting the work that are beyond the control of either contracting party. A contractor's performance is not automatically excused when such events occur. There is no such rule to be found in common law. Contractors assume all risks to performance except for those *force majeure* events that are specifically listed in their contract.

When there is a long list of items and a particular item is not found in the list the court will usually reason that it was the parties' intent to exclude that particular item (Sweet & Schneier, 2013, §17.04E). To avoid exclusion problems, lists are sometimes qualified by adding phrases such as, "and similar events" (Sweet & Schneier, 2013, §17.04E). Yet while those words may expand the scope of the list, they will only expand it to include events that are similar in their meaning to the words that are in the list.

Weather is not ordinarily a *force majeure* event. When it is, it is usually qualified with such phrases as "unusual" weather or "abnormal" weather. Unusual and abnormal are relative terms, however, and one should be cautious about using them. Hurricanes may be common on Okracoke Island, NC but they are unusual and abnormal in Chicago, IL. And while snow and cold winter weather is common in Chicago it is uncommon, although not necessarily unusual and abnormal in eastern North Carolina. A prudent contract would be specific. Maximum sustained winds of over 50-mph is quantifiable by methods endorsed by the U.S. Weather Service. These are less than hurricane force winds but strong enough winds to stop a construction project. Likewise, cold weather can be defined with respect to average daily temperatures, perhaps over some period of consecutive days.

In general, weather conditions may be documented through National Oceanographic and Atmospheric Administration (NOAA) records. But the claim must also provide evidence that the weather actually delayed construction. So for example: ten days and ten nights of rain may be abnormal but if the building were closed in it would be difficult to show that that abnormal weather actually delayed interior construction.

Compensable Delay

Some delays may be compensable. A **compensable delay** is an excusable delay where the payment of additional money to the contractor is also justified. In all cases, the burden is on the contractor to justify both extra time and extra money. Typical causes for compensable delays include:

- errors or omissions in the contract documents;
- concealed or unknown conditions;
- substitute subcontractors caused by the owner;
- change orders;
- construction change directives;
- minor changes to the work;
- failure of payment;
- shut-down, delay, and start-up for hazardous materials;
- emergencies; and
- suspension of the work for the convenience of the owner.

As it is with noncompensable delays, the contractor is generally required to initiate contractual processes by making a compensable delay claim. The designer then decides if an extension and additional compensation are justified by the contract documents and the circumstances of the delay. If the designer determines that they delay and additional compensation are justified he determines the extent of the delay, proposes equitable compensation and prepares an appropriate change order.

Concurrent Delay

Concurrent delay occurs when non-excusable delays and excusable delays overlap. Traditionally, neither party could recover for concurrent delay. The modern trend is apportionment of damages. The main rule of concurrent delay is that if a project is delayed for any reason, no new reason can further extend completion until the delay from the old reason is over. The best way to understand this is through an example.

Example 6.1: Suppose that a commercial project had liquidated damages for late completion of $1,200 per day and the following overlapping delays occur:

- A severe storm shuts the project down from day 15 until day 20;
- A drawing revision to correct errors delays the project from day 19 to 22;
- A major piece of the contractor's equipment goes down from day 21 to 24.

Assume that each of the three delays directly impacted the project's critical path. Figure 6.1 illustrates these delays.

Figure 6.1 Concurrent Delay Example

Solution: Presuming that the storm delay is an excusable delay under the contract's force majeure clause, the contractor gets a 6-day time extension but no extra compensation. The drawing revision is a compensable delay but it cannot extend completion until the storm delay is over, which does not occur until the end of day 20. The equipment failure is a non-excusable delay but it cannot extend completion until the drawing revision delay is over, which does not occur until the end of day 22.

Net result: Substantial completion is delayed 10 days. The contractor receives a total time extension of 8-days, 6-days for the storm and 2-days for the revisions. The 2-day revision delay is also compensable. That compensation to the contractor, however, is offset by a total of $2,400 in liquidated damages for 2-days of equipment delay.

Unless concurrent delays are properly evaluated and resolved they invariably result in disputes that become claims that lead to litigation.

Recovery for Compensable Delay

A contractor may recover actual costs that flow directly from a compensable delay plus a fee based upon a fixed percentage of actual costs. General requirements costs usually flow directly from a compensable delay. General requirements costs are field costs that are necessary for performance of the work but are not connected to any specific work item. Typical costs include rental or lease costs for the jobsite trailers, fencing and security systems, the superintendent's wages and other field management wages. Actual costs, supported by invoices and other documentation, can be recovered on a *per diem* basis for every day of compensable delay.

Under certain conditions, a *pro rata* share of the contractor's home office overhead is also recoverable. Typical home office overhead costs include the rental or lease of the contractor's home office, home office utilities, office supplies, and executive and administrative salaries.

The **Eichleay Formula**, (pronounced; eye'-klee-aye) is a method that is used on federal projects to calculate the amount of a contractor's home office overhead that can be recovered with a compensable delay. The name is derived from the case of *Eichleay v. U.S.* (19xx). Eichleay calculation normally comes at the end of a project when all work has been completed. Recoverable overhead is calculated as shown in Figure 6.2

$$R_E = \left(\frac{B_P}{B_C}\right)\left(\frac{O_C}{D_P}\right) D_E$$

where B_P = total billings for the project
D_P = total days performance for project
D_E = days of compensable delay
B_C = total billings for the company
O_C = total company home office overhead
R_E = recoverable overhead

Figure 6.2 Eichleay Formula

B_C, the total billings for the company and O_C, the total home office overhead for company must be taken over the same period of time as D_P, the total days of performance for the project. The total days of performance for the project include D_E, days of delay.

There are two prerequisites for home office overhead recovery under Eichleay: 1) the contractor must be on standby during a government-caused delay of indefinite duration; and 2) the contractor must be unable to take on other work while on standby. A contractor is able to take on other work while on standby if the contractor actually does take on other work while on standby or if it the contractor had extra work that it could have taken on but chose not to take it on without justifiable cause. A contractor who cannot meet both condition (1) and (2) cannot recover home office overhead under Eichleay.

Acceleration

Acceleration describes an owner's request that the work, or portions of the work, be completed earlier than was contemplated in the contract. Acceleration can be handled through the change process, either by a change order or a construction change directive. Ordinarily, the contractor will want extra money to cover costs incurred to accelerate the work. Disputes on acceleration costs can be avoided if the contractor is consulted and advised before change orders are suggested or construction change directives are issued.

Another type of acceleration, called **constructive acceleration**, occurs when an owner refuses a valid request for a time extension. The owner's refusal has the effect of causing the contractor to accelerate the work, at the contractor's own expense, in order to complete the work on its originally schedule date. Such a forced acceleration is particularly onerous when liquidated damages are also being assessed for late completion.

Constructive acceleration is a common cause for claims. A contractor pursuing a claim for constructive acceleration should be careful, however, because such claims often end up in litigation. The contractor must be able to prove that it was an excusable delay. There must be evidence that timely notice was given to the designer and the owner; that a time extension was actually requested of the designer and owner; that they refused the time extension; and that the owner ordered the project to be completed on its originally contracted schedule. The contractor must have actually accelerated the work. And finally, the contractor must show that extra costs were incurred and that those costs flowed directly from accelerating the work.

No Damages for Delay

A **no-damages-for-delay** clause is an exculpatory clause. An **exculpatory clause** relieves a party from liability arising from negligence or other wrongful acts (Garner, 2006). The no-damages-for-delay clause shields an owner from liability for monetary damages for compensable delays, even when the owner caused the delay. Owners might insist on having them in their contract.

The no-damage-for-delay clause is a one-sided clause that is forced upon a contractor. Despite that, it is a lawful and enforceable clause in most states. Exceptions have been made where it was shown that circumstances occurred that were not within the contemplation of both parties at the time that their contract was formed, when the owner demonstrated bad faith, and when the clause was waived by the words or actions of the owner.

If the contract is negotiable, various strategies can be applied that can mitigate the risks. One of the more fruitful strategies is to enumerate a long list of potential delay circumstances and designate each as a compensatory circumstance or a non-compensatory circumstance. Other mitigating strategies include:

- Establish a ceiling for total dollar exposure;
- Cap the total delay time threshold;
- Recover costs only but no fees;
- Recover specified costs and no fees.

Chapter 7 INTERPRETING THE CONTRACT DOCUMENTS

7.1 Introduction

The contract documents are filled with words, narratives, tabular date and graphic content. The scope and extent of such information yields lengthy and complex documents. The words in these documents are specifically chosen for their special meanings, which may be understood only through particular customs and trade practices within the context of some unique segment of the construction industry. Understanding the contract documents is a challenge. If in the understanding of some element of the contract documents the parties disagree – and disagreements are a pervasive feature of every construction project – they must find some way to resolve their differences. Interpreting the contract documents is the act of arriving at common understanding. The principal rules for interpreting the contract documents are explained in this chapter.

Interpreting the Contract Documents

7.2 Interpretive Evidence

Performing has nothing to do with playing a piano, acting in a play, singing vocals or any of the performing arts. In the construction context, performing has to do with the work of the contract. Performing may be the act of hanging drywall. Hanging drywall is nothing like hanging a picture or hanging out with your friends. It is about permanently fastening sheets of drywall to the framing. The word "sheet" has a particular meaning too, as does "framing." Taping has nothing to do with adhesives and much to do with spreading a very thin skim coat of a gypsum-based plaster on the drywall. A thin strip of paper known as "drywall tape" from which this plastering operation gets its name is but an incidental part of the work of taping.

Courts use to interpret the meaning of the contract documents by the plain meaning of its words (Sweet & Schneier, 2013, §17.04A). To that end, judges (or juries), often lacking meaningful construction experience, were called upon to make common sense interpretations. At best this was an inexact exercise. Words acquire special meaning from the context of their usage and the information known to the parties using the words. Today, courts are likely to consider the special meanings of the words in the contract documents (Sweet & Schneier, 2013, §17.04A).

Courts seek objective means for interpreting the special meanings of words in the contract documents. The testimony from either of the parties is considered too subjective to suit this purpose. Nor can the courts, because of the parol evidence rule, consider any prior written or verbal agreements between the parties. The courts arrive at the objective interpretation that they seek by creating a fiction: the interpretation that would be made by a reasonable person vested with the information known to an informed constructor in the particular context at issue. To that end, the courts look to four types of interpretive evidence: course of performance, course of dealing, usage, and usage of trade.

Course of performance is based upon the premise that the parties themselves know best what they meant by their words. Their actions, because a reasonable person can observe them, are the best objective indicator of what they meant. Course of performance looks to the party's actions pursuant to the contract documents that are at issue. So, for example, if the contract documents instruct the contractor to implement "noise impact mitigation measures" and the contractor had put up all residents living within a 1-mile radius of the site at an oceanfront resort while blasting was being conducted on site, and did so with the owner's knowledge and consent, by the actions of the parties, a reasonable person would interpret "noise impact mitigation measures" as applying within a 1-mile radius of the site and not beyond.

Course of dealing is the same as course of performance except that it looks to how the parties conducted themselves during past projects. Course of dealing looks to the party's actions pursuant to past contracts between the same two parties using the same or similar words as the contract documents that are at issue (*Dept. of Transp. v. 1A Constr.*, 1991). So, for example, if the contract documents instruct the contractor to use "rigid fiberglass insulation, or equal, on concealed ductwork," and on a past project between the same two parties where identical words were used in their contract documents, the contractor had used flexible fiberglass insulation instead, having asserted that flexible insulation has a superior insulating value, and the owner had approved the substitution, a reasonable person would interpret "or equal" to refer to the insulating value and not to the rigidity or flexibility of the insulation (*Cornell v U.S.*, 1967).

Usage is a habitual or customary practice that is implied in the meaning of words (*Lewis v. Jones*, 1952). So, for example, the contract documents call for the owner to pay the contractor $22/hour for labor. The owner balks at paying the contractor for two 15-minute coffee breaks taken by each worker every day of the project. Such breaks are customary and normal. Therefore a reasonable person would interpret labor rates in their common usage, specifically to include 30 minutes per day of non-working time.

Usage of trade is any widely accepted trade custom or practice (*Western States v. U.S.*, 1992). So, for example, the contract documents call for the owner to pay the contractor $60 per square for roofing. The owner measures 960 sq. ft. of roof. The contractor measures the same 960 sq. ft. of roof but adds starter strip and waste in his calculation and then rounds up from 32-1/2 bundles to 11 square. (3 bundles per square where one square = 100 sq. ft. of nominal coverage.) A reasonable person would interpret "square" to mean its usage in the trade, specifically it means all of the roofing that would be needed for the job, in full bundles including sufficient material for starter strips and waste: in other words 11 square.

Lists

When there is a long list of items and a particular item is not found on the list the court will usually reason that it was the parties' intent to exclude that particular item (Sweet & Schneier, 2013, § 17.04E). Let's say, for example, that the contract documents extend the completion date for delays caused by fire, riots, martial law, embargoes, epidemics, strikes, boycotts or pickets. Unusual weather is not on that list. Therefore, it would be specifically excluded from delay claims.

To avoid exclusion problems, lists are sometimes qualified by adding phrases such as, "and similar events" (Sweet & Schneier, 2013, § 17.04E). Yet while those words may expand the scope of the list, they will only expand it to include events that are similar in their meaning to the words that are on the list. Unusual weather would still be excluded from the abovementioned list because none of the events on that list have anything to do with weather. If, on the other hand, the words "unusual weather or any similar events" were appended to the list, the court would include a tornado in this list, even though it was not specifically mentioned on the list, because a tornado is similar to the type of atmospheric events inferred by the word weather. Conversely, the court might not include an earthquake on this list, because an earthquake is not an atmospheric event.

Designer's Authority to Interpret the Design

The contract documents will usually give the designer the authority to interpret the meaning of the design. When interpretation is desired, the contractor prepares a written request, known as a Request for Information (RFI), and the designer responds in writing within a reasonable time or within time limits specified in the contract documents. The designer's interpretation must be reasonably inferable from the contract documents themselves, not from the designer's unexpressed design intentions. In most states a designer has limited immunity from liability for applying his or her design judgment. Nevertheless, nothing precludes a contractor from applying for a change order in the event that the designer's interpretation of the design prejudices the contractor's expectations.

7.3 Harmony

The most important rule of contract interpretation is that the contract will be interpreted as a whole: each and every part of it having meaning (*Downey v. Clauder*, 1992) (*T. Brown v. Pena*, 1997) (*Plaza Dev. Serv. V. Joe Harden*, 1988). The courts seek to harmonize all parts of the contract documents into a coherent whole. No single part can stand-alone. Each part must be read as being part of the whole. When different interpretations are possible, the more harmonious interpretation will be chosen.

How does the court harmonize all parts of the contract documents into a coherent whole? The court does that by applying reason in such a way that the various parts are not in conflict. If they can harmonize the contract documents in that way, then that must be what the parties had intended when they formed their agreement, and that is the interpretation that the court applies. A short example illustrates how this works.

Suppose that the drawings showed 3-1/2" fiberglass insulation and a specification states, "all insulation must be rated R-6." If a contractor can obtain an R-6 rating with 2 inches of insulation then should the contractor install 3-1/2 inches or 2 inches of insulation? To answer that question, the court would look for a way to harmonize what would otherwise appear to be a conflict. To that end, it is clear that no discrepancy between the plans and specifications exists if both 3-1/2" and R-6 express minimum requirements. Therefore, the contractor would be obligated to install at least 3-1/2" of fiberglass insulation. If that insulation happened to be rated R-10 so be it.

Interpreting the Contract Documents

There is no conflict with the specification because R-10 exceeds the minimum rating of R-6. Because the drawings and specification are in harmony when that interpretation is applied, the court would reason that that interpretation is what both parties had intended.

Something called for in the specifications may not be shown on the drawings or vice versa. When that occurs the court will usually harmonize the drawings and specifications by reasoning that what is shown or written on one is implied in the other: that the drawings and specifications work together in this manner.

The contract documents will sometimes include a precedence clause, language saying that the specifications take precedence over the drawings whenever a conflict exists between the two. If your contract has this precedence clause and something was in a specification that did not appear on the drawings your intended precedence rule will not be triggered if your drawings are harmonized with your specifications. Harmony is created between drawings and specifications when what is written or shown in one is implied in the other. The court presumes harmony in every contract and they will go to great lengths to find it, even if the outcomes is not what either party actually intended.

7.4 Conflicting Information

Courts have to find harmony when discrepancies between documents exist. They are not always successful at harmonizing discrepancies between documents. Sometimes there are genuine conflicts between the various written documents of the contract. In that event, the court may look to an order of precedence to determine which document controls (Sweet & Schneier, 2013, §§17.06A-17.06B).

Order of Precedence

There is no common law **order of precedence** among the contract documents. The drawings, for example, do not take precedence over the specifications. The drawings and specifications are equal and they complement each other. Yet though there is no such thing as a common law order of precedence, the common law does recognize a difference between "boilerplate" language and "specific" language.

Boilerplate refers to any document, incorporated into the construction contract that uses standard language. A typical boilerplate is a specification that gets used over and over again on different jobs without any change. When there is a discrepancy between a boilerplate and a specification that has been written especially for a particular project the court will usually determine that the particular language has precedence over the boilerplate language.

In general, specific language has precedence over general language. Custom specifications have precedence over standard (boilerplate) specifications. Likewise, handwritten notes will usually have precedence over printed words (Sweet & Schneier, 2013, §§17.06A-17.06B). The reasoning in these situations is that where the designer took the time to write a handwritten note, what is written must have been what the designer specifically intended.

Some standard form construction contracts include an **order of precedence clause** that sets out which contract document governs in the event of conflict between the contract documents. For example, given conflicts between the drawings and specifications, ConsensusDocs™ 200 §14.2.2 asserts that the specifications govern, and where inconsistency, conflict or ambiguity in the contract documents is found, §14.2.5 asserts the following order of precedence:

a) Change orders and written amendments to the owner-contractor agreement;
b) The owner-contractor agreement;
c) Drawings (large scale governing over small scale), specifications and addenda issued prior to the execution of the owner-contractor agreement or signed by both parties;
d) Information furnished by the owner pursuant to hazardous materials or designated as a contract document at the time of execution of the owner-contractor agreement;
e) Other documents listed in the owner-contractor agreement; and
f) Among categories of documents having the same order of precedence, the term or provision that includes the latest date controls (ConsensusDocs™, 2012).

It bears mentioning here that the AIA standard form construction contracts do not include an order of precedence clause.

Ambiguity

An **ambiguity** is something that is capable of being understood in two or more ways, such as a genuine conflict among the contract documents. Ambiguities in the contract documents are either patent or latent.

A **patent ambiguity** is conflict among the contract documents that is obvious and apparent. If an ambiguity is patent, the contractor has an implied duty, founded in the doctrine good faith and fair dealing, to request clarification. Typically this is done with an RFI. On the failure to seek clarification of patent ambiguity the courts may construe its consequences against the contractor. This duty to seek clarification is also extended to bidding documents: bidders have an implied duty to seek clarification of any patent ambiguities in the bid documents. Failure to seek clarification may preclude the contractor from relief either as to the outcome of the bid or modification to the contract amount after award.

A corollary duty applies to the owner: the owner must provide a timely and meaningful response to a contractor's request for clarification or risk having adverse consequences construed against the owner.

A **latent ambiguity** is a conflict among the contract documents that "…does not readily appear in the language of the [contract documents], but instead arises from a collateral matter when the document[s'] terms are applied or executed" (Patrick, *et al.*, 2010 citing Garner, 2006). Latent ambiguities are basis for compensable delay and other claims during performance of the work of the contract.

Contra Proferentum

When the courts cannot find any reasonable means of resolving a conflict, their last resort may be to apply the rule of *contra proferentum*. Under this rule, the court will construe a contract against the party who wrote the contract. The reasoning behind this rule is that it would be unjust for the party who did not write the contract to be harmed because the party who wrote the contract failed to write words that were not in conflict (Sweet & Schneier, 2013, § 17.04E).

7.5 Misleading Information

The *Spearin* doctrine will not protect a contractor from constructability problems. If what is designed is more difficult or costly to build than the contractor's estimate, the unexpected costs are the contractor's problem. The law will not protect a contractor from bad design practices. The legal principle here is that a contractor enters into a construction contract of his own free will. He is charged with knowing of or finding out about the competence of the owner's designer before entering into the contract. Its just one of the risks that a contractor takes and the law will not protect him from his own poor judgment. But *Spearin* will protect a contractor from being misled by the designer; at least to the extent that misleading plans or specifications result in a costly defect.

Prescriptive Specifications

Prescriptive specifications, also known as **design specifications**, **material and workmanship specifications**, and **method and material specifications**, convey explicit prescriptions that the contractor must comply with in the performance of its work. A typical project will have hundreds of prescriptive specifications, each focused upon the work of a particular trade.

PART 1 - GENERAL		
References	Work included (optional)	Delivery, storage & handling
Submittals	Quality assurance	Sequencing & scheduling
Warranty	System description	Alternates * alternatives
Related work	Project/site conditions	Allowances, unit prices
PART 2 - PRODUCTS		
Materials	Equipment	Acceptable manufacturers
Mixes	Fabrication	
PART 3 - EXECUTION		
Inspection	Schedules	Installation/application/erection
Preparation	Adjusting and cleaning	Extra stock/spare parts
Protection	Field quality control	

Figure 7.1 - CSI 3-part section format

Chapter 8 SUSPENSION AND TERMINATION

8.1 Introduction

The owner-contractor agreement addresses three basic questions: 1) Who are the parties to the contract? 2) How much will be paid to the contractor? and 3) What will be built for the owner? A legally enforceable contract is formed around the answers to just those three questions. All three questions are answered in the owner-contractor agreement. Questions 1 and 2 are answered directly and question 3 is answered indirectly, by enumeration of the contract documents. Thereby, the owner-contractor agreement expresses all of the legally enforceable promises that are the essence of the contract between the parties.

But an enforceable contract is not necessarily a workable contract. Commercial construction projects tend to be large, costly, and complicated. They have a tendency to change, too. Projects must be managed to success. To be effective a manager needs much more than the names of the parties, how much will be paid to the contractor, and what will be built for the owner.

The roles and responsibilities of owner, contractor, designer and subcontractors must be mutually understood. There must be a methodical process for paying the contractor while assuring that the work is performed to the intended quality. There may be a date certain for starting the work, a date certain for completing the work, and perhaps liquidated damages for late completion. Progress must be tracked against a schedule. Insurance will have to be purchased. The owner might want performance and payment bonds. Warranties must be honored. Disruptive events, like severe weather and concealed or unknown conditions, must be anticipated. The contracting parties must maintain their relationship as change happens, delay occurs, or the owner accelerates the project. The parties may want to define how they will manage and settle potential disputes and claims. Last, but not least, the parties will want to define circumstances for suspending work, terminating the work for convenience, or for terminating the work for cause, and the circumstances that define a material breach of the contract between them.

8.2 Conditions

The standard form construction contracts all set out various clauses in a boilerplate document with the name **general conditions**. Some of the clauses in the general conditions are legal conditions. Other clauses are not legal conditions but just instructions, suggestions, or things that the parties are supposed to do to manage their relationships. This collection of legal conditions, instructions, suggestions, and relationship management tools are the rules-of-the-game. The general conditions enable a project to be managed.

Legal Conditions

A promise is a contractual undertaking to perform (or refrain from performing) some designated act. A **legal condition** is a provision of a contract, the fulfillment of which creates (or extinguishes) a duty to perform under the contract (Calamari & Perillo, 1999, Ch. 5.A). An unexcused failure to perform a promise is a breach of contract and gives rise to liability for damages. Failure to satisfy a legal condition, however, is not a breach of contract and does not give rise to liability for damages (Calamari & Perillo, 1999, Ch. 5.A).

A payment clause provides a good example of how legal conditions work. The contractor must present a certificate for payment to the owner. Presentation of the certificate creates a legal duty for the owner to pay the contractor. In the absence of the certificate for payment the owner has no legal duty to pay.

The typical payment process is basically a series of legal conditions. The first condition is usually production by the contractor, soon after contract award, of a schedule of values. The next condition is production by the contractor of an application for payment. Typically the first such request is due after one month of progress. The next condition is approval and issuance of a certificate for payment by the designer, acting on behalf of the owner. The next condition is submittal of the certificate for payment to the owner. The final condition is the tolling of a certain period of time before the owner finally has a duty to pay the contractor. All of the conditions must be satisfied before the owner has a contractual duty to pay the contractor.

Perhaps the most widely used structure for prescriptive specifications is the 3-part section format of the Construction Specification Institute [CSI]. The CSI 3-part section format has the structure illustrated in Figure 7.1.

The contractor must perform the work of the contract in compliance with prescriptive specifications. The *Spearin* doctrine asserts that the owner warrants their adequacy and sufficiency. Contractors who are misled by errors or omissions in the design may use the *Spearin* doctrine as a basis for compensable delay and other claims.

A prescriptive specification cannot be construed as a guarantee of commercial practicability. *Spearin* only warranties performance characteristics. It does not guarantee the availability of supply, timeliness of delivery, the willingness or capability of subcontractors to install the material or other commercial matters.

An exception to the commercial guarantee rule may come into play with a proprietary specification. A **proprietary specification** is a prescriptive specification that calls out a specific manufacturer or supplier. If that manufacturer is unable to supply, there are other suitable manufacturers or suppliers, and the contractor cannot manufacture the material itself, the contractor may be entitled to a compensable delay or a claim for redress.

Some prescriptive specifications may include an **"or equal"** provision or an **"or approved equal"** provision. Both provisions require a contractor to submit proposals for substitute materials or systems to the designer for the designer's approval. The "or equal" provision enables a contractor to freely substitute but the contractor assumes the risk of non-performance of the substitute materials or systems regardless of whether or not the designer approves the substitution. The "or approved equal" provision bars the contractor from installing a substitute material or system without prior approval by the designer. However, designers tend to disclaim performance risks for "or approved equal" substitutions despite having approved their substitution.

The contractor has an implied duty of good faith and fair dealing to report obvious and apparent errors or omissions in specifications to the designer. Failing to notify the designer, the contractor may lose protection under the *Spearin* doctrine.

Performance Specifications

Performance specifications, also known as **output specifications** or **results specifications**, convey certain performance expectations by expressing performance criteria and authorize the contractor to find a technical solution that meets those expectations. Well-drafted performance specifications articulate three components: 1) **Requirements**, qualitative statements of desired performance; 2) **Criterion**, quantitative statements of desired performance; and 3) **Performance Tests**, specific, published evaluative procedures that assure compliance with criterion.

Within the limits of its scope, each performance specification transfers design responsibilities to the contractor. Contractors are not licensed designers, of course, but nothing bars a contractor from subcontracting with firms who are licensed. Contractors engage their responsibilities by delegating design duties to consultants who are licensed in the particular design disciplines that are implicated by a performance specification.

The *Spearin* doctrine is not applicable to performance specifications. However, the contractor will not be held liable if it can be shown that compliance with a performance specification was unreasonably expensive, impractical, or impossible to achieve. Designers should be careful not to overreach with performance specifications wherever they are uncertain about what the market is capable of providing.

As long as complete plans and specifications are delivered to the contractor, the owner provides the implied warranty under the *Spearin* doctrine. However, when a contractor prepares some or all of the plans and specifications the contractor and not the owner provides the warranty (*Leonard v. Home Contractors*, 1916).

Suspension & Termination

Failed Conditions

When a legal condition fails, contracting parties respond in various ways. They can postpone the condition, modify the condition, waive the condition or rescind the condition. Parties can respond to failed conditions in writing but they usually respond by their actions. Waiting longer for the condition to be fulfilled is to postpone the condition. Asking for an alternative thing to be fulfilled is to modify the condition. Proceeding as if the condition was satisfied is to waive the condition. Ignoring the condition is to rescind it. Ignoring the condition is to rescind it. Failure of a condition results in postponement, modification, waiver or rescission of that condition, but failure of a condition can never breach a contract.

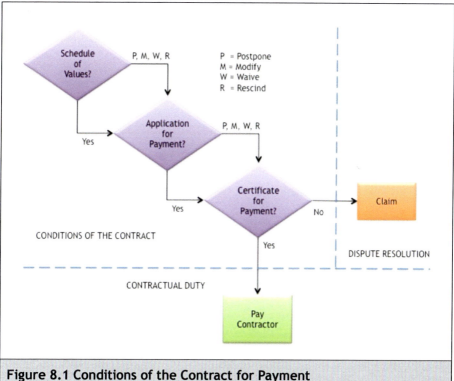

Figure 8.1 Conditions of the Contract for Payment

All of the conditions of the typical payment process must be satisfied before an owner has a contractual duty to pay a contractor. Failure of any one condition would result in postponement, modification, waiver or rescinding of that condition but that failure could never breach the contract,

CONSTRUCTION CONTRACTS

unless one of the parties acted in bad faith. Failure to pay - once the duty arises upon satisfaction of all the conditions preceding that duty - would be a breach of contract. (See Fig. 8.1).

8.3 Breached Promises

There are two ways to breach a contract: a **material breach** or a **minor breach**, sometimes called an **immaterial breach** (Eisenberg, 2002, §§ 816-818). A **minor breach** (or immaterial breach) creates liability for damages in the breaching party. The non-breaching party must continue to perform on the contract. Typically, the non-breaching party makes a claim for damages. If the breaching party refuses to pay the claim, the non-breaching party has the option to **litigate**, meaning a lawsuit to recover damages for breach of contract.

A material breach is evidenced by a substantial failure to perform on a contract (Eisenberg, 2002, §§ 816-818). In the event of a material breach, the non-breaching party is excused from further performance on the contract.

A material breach by an owner typically is the failure of the owner to pay. On an owner's material breach of contract the contractor may stop work, walk away from the contract, and seek to recover actual costs, windup expenses and the expected profit that would have been earned had the project been complete.

A material breach by a contractor typically occurs when a contractor abandons a project. On abandonment the owner may stop paying the contractor, substitute a new contractor, and seek to recover from the breaching contractor any additional costs that arise from substituting the new contractor. As with minor breaches, the non-breaching party has the option to litigate a material breach in order to recover damages.

Material Breach

The court has the task of deciding what **material** means on a case-by-case basis. It is not a straightforward decision. To decide whether or not a breach was material the court must consider the extent to which the breaching party has already performed; whether the breach was willful, negligent, or purely innocent; the extent of uncertainty that the breaching party will perform; the extent that the non-breaching party will obtain the benefits of the bargain; the extent to which the non-breaching party will be compensated; and the degree of hardship imposed on the breaching party (Eisenberg, 2002, § 818). The court's reasoning on material breach is sometimes described as "balancing the equities."

MINOR BREACH	balance of equities	MATERIAL BREACH
	IMPACT OF MATERIAL BREACH ON OWNER	
SEVERE FORFEITURE SMALL PROBLEM SCHEDULE IS RECOVERABLE	MONEY SUBSTITUTE CONTRACTOR COMPLETION	SMALL LOSS THORNY PROBLEM PROBABLY LATE
AT PROJECT START	If OWNER BREACHES	AT PROJECT FINISH
WINDFALL PROFIT SCRAMBLE TO FIND NEXT JOB RISK LOSS OF KEY EMPLOYEES	MONEY SUBSTITUTE WORK WORKFORCE	SMALL GAIN NEXT JOB LINED UP REASSIGN KEY EMPLOYEES
	IMPACT OF MATERIAL BREACH ON CONTRACTOR	

	IMPACT OF MATERIAL BREACH ON OWNER	
LOW RISK OF LOSS SMALL PROBLEM SCHEDULE IS RECOVERABLE	MONEY SUBSTITUTE CONTRACTOR COMPLETION	HIGH RISK OF LOSS THORNY PROBLEM PROBABLY LATE
AT PROJECT START	If CONTRACTOR BREACHES	AT PROJECT FINISH
SMALL LOSS SCRAMBLE TO FIND NEXT JOB RISK LOSS OF KEY EMPLOYEES	MONEY SUBSTITUTE WORK WORKFORCE	SEVERE FORFEITURE NEXT JOB LINED UP REASSIGN KEY EMPLOYEES
	IMPACT OF MATERIAL BREACH ON CONTRACTOR	
MATERIAL BREACH	balance of equities	MINOR BREACH

Figure 8.2 Balance of Equities in Breach Analysis

A court might decide that a particular act was a material breach because it occurred near the end of the project while the same act would have been minor had it occurred near the beginning of the project. Let's take a look at a typical contract breach: an owner does not pay its contractor. Is this a material breach? Or not? In determining whether or not it is a material breach, the court will not penalize the culpable party and reward the aggrieved party. Rather, it will attempt to compensate the aggrieved party for its losses with fairness toward both parties.

From the contractor's perspective a material breach by the owner would result in expectation damages. If this breach occurred early in the project these expectation damages would amount to a windfall profit because little of the work would have been performed. A material breach late in the project would also mean expectation damages but not a windfall because most of the work will have already been performed. From the owner's perspective, she would have to pay all of the contractor's expected profit were her breach a material breach. This would be a very large forfeiture near the beginning of the project but it would be a small forfeiture at the end of the project.

Considering windfall/forfeiture, the balance of equities militates toward declaring this a material breach if it occurs at the end of the project: little is forfeited and there is no windfall for the contractor. The balance of equities militate toward declaring this a minor breach if it occurs at the beginning of the project: the owner pays actual damages but the contractor is not excused from further performance on the contract, he gets no windfall profit and the owner does not suffer a large forfeiture. (See Fig. 8.2)

Also from the contractor's perspective, were a material breach to occur early he would stop performance but then need time to find new work and would risk losing key employees if he were unsuccessful. In contrast, a material breach occurring late would find the contractor with its next job lined up and his key employees already reassigned. From the owner's perspective, when the contractor stops performing as a consequence of her material breach she has to find a substitute contractor. If this happens early in the project her schedule might still be recoverable but if this happens late in the project, her schedule will most likely be delayed because it is difficult to line up a substitute contractor quickly.

Considering hardship/benefits of the bargain, the balance of equities also militates toward declaring this a material breach if it occurs at the end of the project: the contractor will likely receive the benefit of the bargain for terminating near the end of a project while the owner, the culpable party, suffers a small hardship. The balances of equities militate toward declaring this a minor breach if it occurs at the beginning of the project: the owner pays actual damages but the contractor is not excused from further performance on the contract and suffers no hardship. (See Fig. 8.2)

8.4 Suspension and Termination for Cause

Whether any court will hold that particular circumstances are a material breach or a minor breach cannot be known with certainty in advance. The resulting uncertainty is problematic. Let's say that a contractor who is not being paid walks off of the job. Declaring a breach by the owner, this contractor wants to recover expectation damages. This contractor will get what it wants if the court concludes that the owner's failure to pay was a material breach. But if the court concludes that the owner's failure to pay was a minor breach then the tables are turned. A minor breach does not enable a contractor to stop performance on its contract. By walking off the job, the contractor is the one with culpability for material breach whereby the owner becomes the newly aggrieved party. Now the aggrieved owner can recover the losses it incurred in hiring a substitute contractor and the court will render judgment against the now culpable contractor.

In order to mitigate uncertainty most standard form construction contracts define specific legal conditions that, upon fulfilling those conditions, cause a material breach. The courts will enforce material breach in any way that suits the parties, as long as they have expressed it in their contract documents.

Suspension

Suspension by the Owner for Cause

Suspension by the owner for cause authorizes the owner to stop the work in the event of specific actions by the contractor, such as: 1) failing to correct work that does not conform to the contract documents; or 2) repeatedly failing to carry out the work in conformance with the contract documents. The owner must issue a written order to stop work. That stop work order remains valid until the cause for the order is eliminated. A stop work order does not ordinarily preclude an owner from terminating the contractor for cause.

Suspension by the Contractor

> *Over the 7 days between October 6, 2008 and October 14, 2008, the yield on 10-year treasury notes increased by 60 basis points. If you were planning to finance a $10 million dollar construction project this spike could easily cost you an extra $60,000 in interest payments. When interest*

> *rates spike, the owner might just stop paying the contractor. The contractor who isn't getting paid does not want to borrow money just to pay his subcontractors and suppliers especially because interest rates have just spiked up. This contractor wants the contract documents to enable him to suspend the work until the storm of financial uncertainty passes.*

Suspension by the contractor enables a contractor to stop work. Suspension is usually enabled for specific circumstances related to payment failures, such as:

(a) the designer's failure to issue a certificate for payment within a set number of days, typically 7-days, after receipt of the contractor's application for payment; or

(b) the owner's failure to pay the contractor within a set number of days, typically 7-days, of the date for payment set out in the contract documents.

When any of those circumstances occur the contractor must give the designer and owner a set number of days, typically 7-days, written notice before suspending the work. Work can only be suspended until payment is received, at which time the contractor must start-up again. The contractor can usually recover costs for shutdown, delay, and startup. Interest and other costs may also be recovered if they are stipulated in the contract documents.

Termination

Termination by the Contractor

> *Sixty days have gone by, the work is suspended and the owner has still not made payment. This contractor wants to terminate the contract, cover his costs and overhead, collect the profit that he would have earned had he completed the work, and move on to the next job. To assure this the contract documents must enable termination by the contractor for payment failure.*

SUSPENSION & TERMINATION

Termination by the contractor enables a contractor who is not at fault to terminate its contract when a work stoppage of a set number of days, typically 30 consecutive days, was caused by specific reasons stated in the contract documents, such as:

(a) A court order requiring all work to be stopped;
(b) A declaration war or other emergency issued by the government;
(c) The designer, without justifiable cause, has neither issued a certificate for payment nor has notified the contractor of the reason for withholding certification; or
(d) The owner has failed, upon request by the contractor, to provide information to enable the contractor to enforce its mechanics' lien rights.

One of the reasons stated in the contract documents must apply for the contractor to terminate. When the contractor is entitled to terminate, the contractor must give a certain number of days, typically 7 days, written notice to the designer and owner (in addition to the set number of days of suspension, typically 30 days) before terminating. If the suspension had occurred because the contractor stopped work for non-payment, the contractor must still wait for the set number of days of suspension, typically 30 days, unpaid and with work suspended, and then provide an additional number of days, typically 7 days, written notice to the owner and designer before terminating.

The contract documents may allow termination for repeated suspensions, delays or interruptions of the work due to suspensions by the owner for convenience that are aggregated over a longer period of time, perhaps as much as one year. Upon fulfilling these conditions an additional number of days, typically 7 days, written notice must be given to the designer and owner before terminating.

The contract documents may allow termination because the owner repeatedly failed to fulfill its obligation and this failure impacted the progress of the work so as to cause a work stoppage of a specific number of days, typically 60 consecutive days. Again, upon fulfilling these conditions an additional number of days, typically 7 days, written notice must be given to the designer and owner - in addition to the set number of days of suspension - before terminating.

If the contractor, a subcontractor, sub-subcontractor, their agents, employees, or any other persons or entities other than the owner is responsible for the work stoppage, the contractor cannot terminate the owner-contractor agreement.

Termination by the Owner for Cause

An owner has been paying the contractor on time, each and every month, but the contractor has not paid any of his subcontractors for months. This owner fears that if the subcontractors start to file mechanics' liens the banks will pull his construction financing. The owner wants to terminate this contractor, get its subcontractors paid, and hire a substitute contractor. To assure this the contract documents must enable termination by the owner for cause.

Termination by the owner for cause enables an owner to terminate its contract for reasons such as when the contractor:

(a) Repeatedly refuses or fails to supply enough properly skilled workers or proper materials;
(b) Fails to make payment to subcontractors for materials or labor in accordance with the respective agreements between the contractor and its subcontractors;
(c) Repeatedly disregards applicable laws, statutes, ordinances, codes, rules and regulations, or lawful orders of a public authority; or
(d) Otherwise is guilty of substantial breach of a provision of the contract documents.

Wrongful conduct must occur repeatedly. Instances of insufficient numbers of workers or improper materials must also occur repeatedly. Isolated infractions will not justify termination.

Termination by the owner for cause is not automatic. When any one of the listed causes, and only those causes, the designer or other authorized decision maker, must decide if the infractions are sufficient to terminate the contract with the contractor. The owner and contractor may choose to name a third-party, not the designer, as the authorized decision maker. The name of this third-party decision maker would be inserted in the owner-contractor agreement.

SUSPENSION & TERMINATION

On a decision to terminate the authorized decision maker would have to provide a set number of days, typically 7 days, written notice of termination to the contractor and the surety, if any. Termination ordinarily enables the owner to stop payment to the contractor until the work is finished. A typical termination clause enables the owner to do the following:

1. Exclude the contractor from the site and take possession of all materials, equipment, tools, and construction equipment and machinery thereon owned by the contractor;
2. Accept assignment of subcontracts; and
3. Finish the work by any reasonable method.

Upon written request of the contractor, the owner would ordinarily furnish a detailed accounting of the costs incurred by the owner in finishing the work. The contractor may be liable for monetary damages incurred by the owner in completing the work after termination of the contractor. If monetary damages exceed the unpaid balance of the contract the contractor has to pay the difference. If the contract documents include a consequential damage waiver it may limit the amount of monetary damages that may be recovered. The authorized decision maker will ordinarily do a final accounting to determine if any money is due to the contractor. The contractor must make an application for payment after the work is finally complete for certification by the authorized decision maker.

If there is a performance bond the surety has the legal right to step into the contractor's shoes and either complete the work or hire a substitute contractor to finish the work. The owner should consult with their legal counsel before taking action whenever a surety is providing either performance or payment bonds.

Upon termination for cause, all of the terminated contractor's subcontracts might be assigned to the owner. On such assignment, the owner would acquire all of the rights and duties of the terminated contractor with respect to subcontractors. This enables the owner to step into the shoes of the terminated contractor and continue the work. It also obligates the owner to pay any amounts that were due the subcontractors at the time of the contractor's termination. There could be reasonable upward adjustment of subcontract prices if the work had been suspended for more than 30 days. But for this one exception, subcontractors are not entitled to change their prices just because there has been a termination followed by an assignment of their contract.

Should the owner choose to hire a substitute contractor the owner would reassign all of the subcontracts to that substitute. Payment obligations cannot be reassigned, however. So the owner will remain liable for payments to the subcontractors. The owner should consult with legal counsel because there may be legal issues associated with assignment. The sureties of the subcontractors should also be contacted because there are bonding issues associated with assignment.

8.5 Suspension or Termination for Convenience

Suspension by the Owner for Convenience

Over the 7 days between October 6, 2008 and October 14, 2008, the yield on 10-year treasury notes increased by 60 basis points. If you were planning to finance a $10 million dollar construction project this spike could easily cost you an extra $60,000 in interest payments. Yet only two months later, on December 18, yields had fallen by 200 basis points. Waiting until December to finance construction would have saved you $160,000 in interest payments. Sometimes an owner will want to suspend work on construction while waiting for a storm of financial uncertainty to pass. To assure this the contract documents must enable suspension for convenience.

The contract documents may enable the owner to suspend, delay or interrupt the work. Notice to the contractor must be in writing. The owner need not show cause for suspension. Work can be suspended for good reason or for no reason at all. Most convenience clauses place some limitation on the owner, however, by giving the contractor grounds to terminate the contract for repeated suspensions, delays or interruptions of the work by the owner. The contractor is ordinarily entitled to additional money and time as a consequence of suspension by the owner.

Termination by the Owner for Convenience

Buildings under construction are vulnerable to fire. When work in progress is destroyed by fire the owner just might want to use the insurance proceeds to clear away the rubble, return the site to green field conditions, and call it a day. To assure this the contract documents must enable termination by the owner for convenience.

The contract documents may enable the owner to terminate its contract for any good reason or no reason whatsoever. Convenience termination is enforceable in most states. Such clauses ordinarily entitle a contractor to actual costs, reasonable overhead and profit.

8.6 Delay and its Implication on Material Breach

Acts by either party that cause delays are less likely to be construed as material breaches than are construction shortcomings by the contractor or payment failures by the owner (Sweet, 1997, §9.8). A contractor's delayed performance is excusable were it caused by design errors or omissions, changes, or suspension of the work by the owner. A contractor's delayed performance is unexcused were it caused by any other act not in the control of the owner. Remember, the contractor assumes all risks of performance, including events causing delay, unless otherwise excused in the contract.

Proving damages for unexcused delay is very difficult to do (Sweet, 1997, § 9.10(c)). To prove delay an event either in the control of the owner or contractually excused must be directly linked to the delay. The link is established by probing scheduling information. Contractors, depending on their sophistication, might have simple bar chart schedules, computer-generated Gantt chars, critical path (CPM) schedules, or excruciatingly complex PERT/CPM schedules.

The standard form construction contracts do a comprehensive job of defining delays: the various events that constitute excusable delays (delays caused by an event that is either in the direct control of the owner or expressly excused in the contract); compensable delay (excusable delays that award extra time and money to the contractor); noncompensable delays (excusable delays that award extra time but no extra money to the contractor); and inexcusable delays. But when the contract is silent about delay, the contractor is generally at risk and therefore extra time or money is beyond his reach unless it can be shown that the delay was caused by something clearly in the control of the owner or the owner's designer(s).

8.7 Consequential Damages

Consequential damages are secondary cost effects that occur after actual damages are incurred. They do not flow directly from the alleged breach, but are an indirect source or a result of such breach (Patrick, et al., 2010 citing Kelleher, Abernathy & Bell, 2008). An owner, for example, may claim consequential damages for rental expenses, losses of use, income, profit, financing, business and reputation, and for loss of management or employee productivity and services. A contractor may, for example, claim consequential damages for principal office expenses including the compensation of personnel stationed there, for losses of financing, business and reputation, and for loss of profit except anticipated profit arising directly from the work.

Common law restricts consequential damages to damages that the aggrieved party would have reason to anticipate at the time that their contract was formed. So if, for example, the contract documents for a commercial office building call out LEED platinum certification, the contractor should anticipate that the owner would lose significant income if that certification were denied. If, in fact, certification is denied with the consequence that a retailer of green products cancels its lease, this lost rental could be a valid, consequential damages claim.

Construction contracts increasingly include consequential damage waiver clauses whereby both parties are barred from seeking consequential damages from the other party. The parties may elect to insert a mutual waiver of claims for consequential damages in their contract. This waiver bars either party from making consequential damages claims against the other. A waiver of consequential damages makes a no-damages-for-delay clause less meaningful.

For a consequential damages waiver to be fully effective, however, contractors should flow the waiver language down into their contracts with subcontractors and suppliers and so forth. Likewise, the owner should insert a mutual waiver of claims for consequential damages within the owner-architect agreement.

Chapter 9 ASSURANCE

9.1 Introduction

Assurance is a pledge, warrantee or guarantee that gives confidence or security (Garner, 2006). Construction owners find some assurance in warranties provided by their contractor. A **warranty** is a certification that certain aspects of a project are as promised, effective for a certain period of time. A warranty may be expressed in writing. This is called an **express warranty**. A warranty may also arise from law. This is called an **implied warranty**.

Assurrance

Owners find additional assurance in the periodic inspections of the jobsite that they conduct to stay informed of work progress, conformance with the contract documents and to alert them to potential construction defects. Testing may also be performed to assure that quality standards are being met. The contractor may be restrained from covering up certain categories of work until the owner has an opportunity to inspect. Work that is prematurely covered may have to be uncovered and recovered at the contractor's expense.

Construction defects that are discovered either during construction or after construction, sometimes long after final completion, give rise to liability in the contractor for the cost of correcting these defects.

This chapter explains the legal principles of assurance and describes the general conditions that establish express warranties, and rules for inspection, testing, uncovering, and correcting defects during and after construction.

9.2 Warranty

The contract documents will include an **express warranty** as to the quality of the contractor's work. Such a warranty ordinarily has three elements:

1) Materials and equipment will be new and of good quality;
2) The work will conform to the contract documents; and
3) The work will be free from defects.

The contract documents can create exceptions, such as specifying the reuse of old materials. Damage or defects caused by abuse, alterations, improper or insufficient maintenance, improper operation, normal wear and tear and normal usage during operation are often excluded in an express warranty.

The courts have interpreted express warranties to place responsibility for a quality problem on the party responsible for the problem. This is a simple concept but it proves to be very difficult to apply because it is seldom clear whether the owner, contractor, or designer should be held responsible. There are some general rules, however, that are helpful.

If the problem arose because of the design, either the designer or the owner will normally be held responsible. If the designer is held responsible the designer will have to bear the cost of changing the plans but the owner is responsible for any increased construction costs. The designer can avoid responsibility for the cost of changing the plans by demonstrating that their design work was performed as competently as any other designer, similarly situated, would perform their work. Design is an imperfect art. The owner is held responsible for design imperfections for the reason that the owner selected the designer and had the power to control his or her work. You cannot escape your own business decisions.

If the quality problem arose because of the way that something was constructed then the contractor is usually held responsible. An exception would occur, making the owner responsible instead, if the quality problem were inherent in the way that the work was required to be constructed. The presumptions here are that:

1) The contractor did exactly what the contractor was instructed to do in the contract documents; and
2) Any inherent quality problems that the contractor knew of (or should have known of) were reported to the designer, who then

instructed the contractor to proceed as directed in the contract documents.

Material and equipment quality problems can be categorized as improperly installed, improperly specified, or improperly manufactured. Improperly installed materials are usually going to be the responsibility of the contractor. Material and equipment is improperly specified when it is unsuitable for its design purpose. Material and equipment is improperly manufactured when it is suitable for its design purpose except for quality problems that arise during manufacturing.

Responsibility for materials and equipment that are improperly specified or improperly manufactured can go either way depending on how the specifications were written. The owner - because he or she has power to control the architect - will usually be responsible for materials and equipment on prescriptive specifications, especially prescriptive specifications that require the contractor to purchase specific materials or equipment models. Conversely, the contractor will usually be responsible for materials and equipment on performance specifications.

Warranty and Material Breach

As described earlier, most people understand a warranty to have effect when something is found defective, after construction is complete. It is that. But a warranty can also have effect during construction. The implied warranty of workmanlike performance is such a warranty. A contractor who persistently fails to supply suitably skilled workers could suffer the consequences of breaching his implied warranty of workmanlike performance. The consequence of that breach could be early termination of the contractor's contract plus the award of damages to the owner. Damages in this circumstance are usually in the amount of any difference in price for having a different contractor finish the work.

With most contractual disputes the party claiming a breach has the burden of proving that the breach was material. Not so with breach of warranty during construction. Failure to meet a warranty obligation before substantial completion occurs is conclusively presumed to be a material breach of contract. Conclusively presumed is a legal phrase that, in this context, means that the material breach of contract is automatic. It is not necessary for the party claiming the breach to prove that the breach was material. It is only necessary to show that the warranty obligation was not met. However, any breach of warranty occurring after substantial completion is not a material breach. There are no material breaches after substantial completion. A breach of warranty after substantial completion still gives rise to liability, however.

Construction Defects

The words patent defect and latent defect have legal significance. A **patent defect** is a significant act or omission in the contractor's performance that clearly presents itself in the course of an ordinary inspection. Common law will not hold a contractor liable for patent acts, omissions, or defects that are discovered after substantial completion. This comes about under the presumption that an inspection was or should have been performed at the time of substantial completion. In short, the owner either knew or should have known.

The opposite is generally true for latent acts, omissions, or defects that are discovered after substantial completion. A **latent defect** is a significant act or omission in the contractor's performance that conceals itself from discovery in the course of an ordinary inspection. Common law will hold a contractor liable for latent acts, omissions, or defects that are discovered after substantial completion. This comes about under the presumption that a diligent inspection at the time of substantial completion would not have revealed a latent defect to the owner.

9.3 Inspection

The owner has a right but generally not the obligation to inspect the work of the contractor. The substantial completion inspection and the final completion inspection are exceptions to that rule. The contract documents may require the designer to perform those inspections: which the designer does on behalf of the owner, of course. The rights and remedies that the owner has with respect to warranties and construction defects can be lost for deficient substantial completion or final completion inspections.

Whatever it is that the owner chooses to inspect, the purpose of that inspection will be to protect the owner, not to protect the contractor. Owners will also want someone to make periodic visits to the site. The designer usually makes these periodic visits - acting in its capacity as agent for the owner - because the designer is presumably best suited to evaluate progress of the work, the work quality, and to identify any construction defects. The designer keeps the owner informed of work progress and conformance with the contract documents and alerts the owner to potential construction defects.

If, upon conducting an inspection, nonconforming work is discovered, the designer may choose to:

1. Accept the nonconforming work;
2. Require correction of the nonconforming work;
3. Reject the nonconforming work and have it torn out and replaced; or
4. If appropriate, terminate the construction contract for cause.

The contractor is vested with the primary responsibility for performing the work in conformance with the contract documents. To that end the contractor cannot rely on the owner's inspections. The contractor must monitor its own work and the work of its subcontractors and suppliers. Comprehensive inspection programs, often engaging independent third-party inspectors, are executed in the fulfillment of the contractor's responsibilities. A comprehensive inspection program should include regular and continuous inspections of the work in place, job conditions, and progress monitoring against schedule.

Federal, state and local governments and their agencies may rely on their contractor's quality assurance system entirely or just for specific inspections and tests that are specified in the contract documents. Many standard sharing clauses are utilized for that purpose. Public agencies may also stipulate higher-level quality requirements. These higher-level quality programs operate separate and apart from their contractors' quality assurance programs.

Inspection of Goods

The Uniform Commercial Code [UCC] governs contracts for the sale of goods. Goods must be inspected and accepted before they can be incorporated into the work. Inspection and acceptance intersects a project's quality assurance process. The owner may exercise indirect influence over the goods inspection process.

UCC §2-105, defines goods as "...things which are movable at the time of identification to the contract for sale." Building materials are moveable. So they fit the description for goods. Neither the building nor the land under it is moveable though. Therefore, buildings and land do not fit the description for goods. Once the building materials get incorporated into or attached to the building they are no longer moveable. Therefore, they lose their distinction as goods. Things that are not physical objects are not goods either. Services are not physical objects. Therefore, services do not fit the description for goods.

Receipt of materials on site does not confer acceptance. UCC §2-513 gives the contractor the right to inspect the materials before accepting them. The inspection must be completed within a reasonable time. The meaning of reasonable depends on the circumstances. Concrete will be tested and inspected while the ready mix truck is still on site. Cubes of bricks can be inspected the same day. Complex machinery may take several days to inspect because there are protective crates to remove and lots of parts and pieces to count.

Compressive strength tests or other destructive sampling is authorized by UCC §2-513(1) wherever such tests are normal and customary. A reasonable amount of materials can be consumed in testing and sampling without prejudice to the contractor. Destructive tests must be completed within a reasonable amount of time after receiving the materials.

The parties can stipulate in their contract that the inspection will be at the supplier's shop before shipment or at any other location at whatever time suits them. Such stipulations will be enforced. The contractor must pay all costs for inspections, even those performed at a supplier's shop. But if the inspection reveals nonconforming goods and the supplier replaces those goods, the supplier has to pay for subsequent inspections.

Constructive Change

Constructive change arises from circumstances that have the effect of requiring the contractor to perform extra work as if a change order existed, but where a formal change order does not in fact exist. The origins of constructive change are founded in misinterpretation of the contract documents, improperly performed inspections, owner denials of time extensions for excusable delay, minor changes in the work, construction change directives, and defective drawings or specifications. Common sources of constructive change include erroneously higher standards of performance, erroneous rejection of acceptable work, and wrongful delay and disruptions

The imposition of an erroneously higher standard of performance is associated with the testing of materials and systems. Tests should not be more stringent than contemplated in the contract documents. Nor should they be inconsistent with industry practice. The cost to a contractor of an erroneously higher standard of performance could be construed as a constructive change.

Erroneous rejection of acceptable work occurs because an inspector who has contractual authority to accept work has a corollary and implied authority to reject work. The inspector has no authority to change the contract, of course, but could wrongfully reject acceptable work. The cost to the contractor of an erroneous rejection could be construed as a constructive change.

Wrongful delay and disruption occurs when the performance of the work is hindered without justifiable cause. A designer's failure to complete submittal reviews within a reasonable time is a common cause of wrongful delay and disruption. Provided that the contractor has submitted a viable submittal schedule the designer must timely perform its review and approvals. Excessive and untimely inspections by the owner are another common source of wrongful delay and disruption.

A contractor who suffers a constructive change should promptly notify the owner and designer. The contractor may continue to perform under protest but would be ill advised to suspend or terminate work in reaction to a presumed constructive change.

9.4 Correction of Work

The contract documents will set out rules for correcting work that does not meet the contractor's express or implied quality warranties. This work is known as **warranty work**. The contractor must correct warranty work at its own expense.

The contract documents may flag work that has to be examined by the designer before it is covered up. If that work is prematurely covered up, the contractor will have to uncover and recover at his own expense. The designer can also request uncovering work that was not flagged for examination in the contract documents. When that happens the contractor may have to pay the cost to uncover and recover if what is uncovered does not conform to the contract documents. Conversely, the owner has to pay if what is uncovered does conform.

Statutes of Limitations

Some of the standard form contracts require the contractor to correct both latent and patent defects, even those discovered well after substantial completion. A contractor who faces perpetual liability for defects will mitigate its risk by adding a risk premium to its bid. This, of course, is just another cost of doing business that gets passed through to the owners. Owners, particularly the government and government agencies, have noticed. During periods of rising construction costs, dampened construction spending, and low construction employment, legislators in the various states have stepped in to enact statutes of limitations that cut off contractors' liabilities for patent and latent defects after a period of years.

California, for example, regulates the liability period for patent defects discovered after substantial completion (California Code of Civil Procedure [Cal CCP], 1967). Lawsuits for injury to property, persons or wrongful death arising from patent defects must be filed within four years after substantial completion. An exception is made for injury or death occurring during the fourth year after substantial completion. Lawsuits for these exceptions may be brought to court within one year of the date of the occurrence but not more than five years after substantial completion.

California also regulates the liability period for some latent defects discovered after substantial completion (Cal CCP, 1981). Lawsuits for injury to property arising from latent defects must be filed within ten years after substantial completion. An exception is made for injury to property based on willful misconduct or fraudulent concealment. There is no time limit on suits for injury or death arising from latent defects until the injury or death occurs but these lawsuits must be brought to court within one year of the date of injury of death.

The statutes of limitation on patent and latent defects in California apply to all types of construction except for some residential construction. California has adopted extensive rules covering residential construction defects applying to new, owner-occupied, single family homes (or condominium units) built since January 1, 2003 (California Civil Code [Cal CIV], 2002).

One-Year Correction Period

Some standard form construction contracts create a **one-year correction period** for warranty work. If warranty work is discovered during this correction period the owner has a contractual duty to notify the contractor in writing. Upon notification the contractor then has the right to fix the work at the contractor's own expense. If the contractor does not fix the work within a reasonable time after notification the owner may fix it and the contractor is obligated to reimburse the owner's costs. Reimbursable costs include uncovering and recovering the work. If the owner fixes warranty work during the correction period without first notifying the contractor the owner does the work at the owner's own expense and waives his right to make a claim for breach of warranty.

The one-year correction period usually starts on the date of substantial completion. If warranties start on a date other than substantial completion date the one-year correction period begins on that date. If any portion of the scope of work was first performed after substantial completion the correction period starts on the actual date of completion of that portion.

If warranty work is discovered after the one-year correction period the owner may fix it without notifying the contractor and the contractor is obligated to reimburse the owner's costs including the cost of uncovering and recovering the work. If warranty work was fixed by the contractor but needs further correction more than one year after substantial completion the owner may fix it without notifying the contractor. The owner who fixes a contractor's warranty work after the one-year correction period may file a claim against the contractor or litigate to recover incurred costs.

Contractors ordinarily have to remove nonconforming work that is neither corrected nor accepted by the owner. A contractor will have to fix, at its own expense, any work of other constructors that it damages. And, the owner usually has the right to accept nonconforming work and authorize a deductive change order as compensation for lost value.

Chapter 10 CLAIMS AND DISPUTE RESOLUTION

10.1 Introduction

AIA A201-2007 §15.1.1 defines a **claim** as a demand or assertion by one of the parties seeking payment of money, or other relief with respect to the terms of the contract. A claim also includes other disputes and matters in question between the owner and contractor arising out of or relating to the contract documents (American Institute of Architects [AIA], 2007b).

Claims and Dispute Resolution

Most construction claims are resolved through negotiations between the owner's representative and the contractor's representative. But when negotiations fail, it triggers the particular dispute resolution processes, if any, that are stipulated in their contract documents. If the contract documents are silent about dispute resolution processes the parties can always litigate their dispute. Litigation is a lawsuit for damages. It is the traditional way of resolving a dispute. However, most construction disputes are settled by an **alternative dispute resolution [ADR]** method, the most common of which are mediation and arbitration.

10.2 Claims

Claims can be categorized under three different types of legal actions: breach of contract, breach of warranty, or torts. All such legal actions must be filed within the applicable statute of limitations in the state where the work is being performed. The contract documents may set a maximum time limit, such as 10 years, for initiating binding dispute resolution but that time limit can be reduced by the applicable statute of limitations in the state where the work is being performed.

A **claim document** consists of a recitation of all pertinent information that is needed to describe, evaluate and settle a claim. The main body of the claim document is a factual narrative. The factual narrative tells the story. The claim document should include a recapitulation of all amounts claimed, a detailed cost breakdown, and copies of all invoices and other documents to support those costs. The strongest documentary evidence should be summarized. **Documentary evidence** consists of project records, memoranda, daily logs, computer files, photographs, and other such things. The claim document is also a showcase of persuasive demonstrative evidence. **Demonstrative evidence** includes such things as bar charts, pie charts, scatter plots, graphs, Gantt charts, pert diagrams, s-curves, resource loading diagrams, photographs, animations, simulations and videos.

Demonstrative evidence is much more persuasive than the tedious recitation of facts. It helps to simplify and clarify complex issues. It may be the most important part of the claim document. The contractor is, after all, trying to persuade the owner and designer that its claim is justified and thereby avoid the expense of litigation. Nevertheless, the claim document should include a discussion of applicable legal principles. Sometimes a convincing explanation of the likely outcome of litigation is enough to motivate an owner to settle a claim.

The responsibility to substantiate a claim rests with the party making the claim, who then initiates the claims process by sending a written notice to the other party. The claim document need not be provided upon initiation of the claim. Initiation must occur, however, within a set number of days, typically 21 days, of the event or condition causing the claim. The contract documents will usually stipulate that the work must continue after initiating a claim to avoid the wasteful expense of suspending and restarting the work each time a claim is processed. There may be circumstances, however, where suspension or termination is actually justified.

The Initial Decision

Construction contracts in continental Europe typically give decision-making powers to an independent third-party. The designer is not given the authority to resolve disputes, not even the authority to make a decision subject to review or the opportunity to make a non-binding initial decision (Sweet, 1997, p. 38). The reasons in support of this are:

- The decision-maker must make impartial decisions that respect the interests of all parties;
- The designer may have a fiduciary duty to do what is in the best interest of the owner;
- The designer cannot make impartial decisions because the designer is the fiduciary agent of the owner.

Designers in continental Europe are, however, encouraged to express their opinions on interpretation of the contract documents.

Construction contracts in the USA typically give broad decision-making powers to the designer. A number of reasons are generally given in support of this:

- Stature and integrity is associated with the architecture and engineering professions. There is a general sense that these professionals behave ethically and fairly.
- The architect or engineer created the design; therefore, he or she is the best-informed interpreter of the meaning and intent of the design.
- The architect or engineer's knowledge is superior to that of the owner; therefore, as they provide professional advice that protects the owner and champions the owner's best interests.
- Any alternative to the design professional as decision-maker would be worse for the owner.
- A quick decision by the architect or engineer, even if it is the wrong decision, is better than a decision by a third-party that would take much longer to make.

Starting with the 2007 edition of A201 the AIA enabled the parties to name a third party who is not the architect to be the decision-maker (AIA, 2007b). For that purpose a new identity, the **initial decision maker,** was created. The AIA presumes that the architect will serve as the initial decision maker for changes, disputes, and claims. But filling this role with an independent third party reflects the decision-making strategy long in place in continental Europe. Unfortunately, A201 does not explain how the initial decision maker is selected, how he or she is paid for, whether or not the initial decision maker is licensed, or carries professional liability insurance (AIA, 2007b).

The initial decision maker has two primary roles. First, he or she makes an initial decision on claims by either the owner or the contractor. Second, he or she makes a determination if grounds exist for the owner to terminate the contractor for cause.

The party making a claim must present its claim to the initial decision maker first, for an initial decision on the claim. Claims arising from hazardous materials, emergencies, and the owner's distribution of proceeds from insured losses are exempted. These exempt claims may proceed directly to mediation. The initial decision maker should solve most claims, thereby avoiding the added expense of more formal dispute resolution processes.

The initial decision maker has a set number of days, typically ten days, to take one or more of the following actions:

- Request additional supporting data from the other party,
- Reject the claim in whole or in part,
- Approve the claim,
- Suggest a compromise, or
- Advise the parties that the initial decision maker is unable to resolve the claim.

If either party disagrees with the initial decision maker the claim survives to trigger the dispute resolution processes, if any, that are stipulated in their contract documents. The contractor should contact the surety at this stage because the surety has a legal right to and likely a willingness to step in and help the contractor to resolve the problem. Surety intervention is often in the best interest of all parties involved.

10.3 Dispute Resolution

The construction industry has been a leading innovator in ADR. Non-binding mediation may be the most popular form of ADR within the construction industry (Stipanovich, 1996, p. 10). Binding arbitration is the most widely used form. (Stipanovich, 1996, p. 13). Other non-binding ADR processes are also emerging, including dispute review boards, the project neutral, the mini-trial and the summary jury trial. Litigation, the traditional means of resolving disputes, still remains the default means of binding dispute resolution. The standard form construction contracts vary in the ADR processes that they embrace.

EJCDC C-700 Standard General Conditions of the Construction Contract stipulates that the engineer of record make an initial administrative decision (Engineers Joint Contract Document Committee [EJCDC], 2007). If either party is unhappy with the engineer's decision they may take it to non-binding mediation. If the mediation does not result in a settlement the parties have 30 days to invoke a dispute resolution process or the engineer's original decision becomes final. The parties may, by mutual agreement, use any dispute resolution process; or a dispute resolution that they had chosen by prior agreement in their contract documents; or take their claim to litigation.

ConsensusDocs™ 200 Standard Agreement and General Conditions Between Owner and Constructor have the representatives of the parties enter into direct discussions (ConsensusDocs™, 2012). Should discussions fail, the claim moves on to mediation or mitigation. Two nonbinding dispute mitigation methods are possible, a project neutral or a dispute review board. If the claim cannot be resolved through mediation or mitigation, the claim gets resolved through binding dispute resolution. Two binding dispute resolution methods are possible, litigation or arbitration. The choice of mediation or mitigation; the type of mitigation, either a project neutral or a dispute review board; and the choice of binding dispute resolution methods, either litigation or arbitration, are by prior selection at the time of signing the ConsensusDocs™ agreement (ConsensusDocs™, 2012).

AIA A201™-2007 General Conditions of the Contract for Construction direct the disputing parties to go through non-binding mediation (AIA, 2007b). If a settlement cannot be reached through mediation, the parties may proceed to arbitration or whatever other alternate dispute resolution method had been specified in their contract. The AIA's default position is now litigation (AIA, 2007b). Regardless of the chosen dispute resolution method, the parties must go through mediation first. They cannot bypass mediation and go directly to binding dispute resolution.

Dispute Review Boards

Dispute review boards are widely used for public construction projects. The boards usually consist of three members. Typically, the owner selects one board member, the contractor selects a second board member, and then those two members select the third board member.

Unlike with mediation or arbitration hearings, dispute review boards are formed before there are any disputes between the contracting parties. They are formed at the beginning of the project and they meet at regular intervals during the project. The dispute review board meeting is a forum for the key project managers to report on the status of the work. Because there are regular meetings, issues can be discussed in a timely way and actions can be taken to prevent issues from degenerating into claims and disputes. If any issue does evolve into a dispute, the dispute review board is the first decision-maker for that dispute.

Decisions of dispute review boards are not binding on the parties. Anecdotal evidence suggests that dispute review board decisions are generally accepted, however. That may be because all parties have some sense of control over the process. It is important for the board members to be independent of both owner and contractor, so that their impartiality cannot be questioned. Perhaps the greatest benefit that comes from dispute review boards is the regular communications that they occasion and that disputes can be avoided or settled without undo interruption to the work in progress.

Project Neutral

A **project neutral** resembles a dispute review board. It is formed at the beginning of the project, holds regular meetings, improves communication and avoids or settles disputes before they become disruptive to the project. It is usually a group of construction experts as opposed to the three-person dispute review board. Their expertise provides more satisfactory results but at much higher cost.

Some projects have experimented with a single expert. This expert is generally referred to as a **standing neutral.** A standing neutral provides many of the same benefits as a project neutral or dispute review board, but at much lower cost.

Mini-Trial

The term **mini-trial** is a misnomer. It is not a trial at all. Mini-trials are formed to resolve disputes. They are an efficient, low cost substitute for other forms of non-binding dispute resolution. A mediator conducts a mini-trial: preferably one who has expertise in construction contracting. The unique feature of mini-trials is that senior executives of the disputing party are the focus. Because senior executives are not usually involved in day-to-day project issues their positions are not as battle-hardened as those of the direct project participants. More importantly, senior executives possess the authority and often the eagerness to settle. In the mini-trial both sides present their case to the company executives while the mediator, acting as a facilitator, tries to find grounds for settlement. Although not in widespread use, mini-trials have had some notable successes.

Summary Jury Trial

A **summary jury trial** is an invention of the federal court system. In a summary jury trial, the lawyers present combined opening and closing arguments in front of a jury. This usually takes a day or less. The juries are then instructed to render a verdict.

Summary jury trial verdicts are non-binding. Conducted before the real trial begins, summary jury trials provide the disputants with a realistic appraisal of the strengths and weaknesses of their cases. Their purpose is to motivate the parties to settle. A summary jury trial may extinguish the need for arbitration or litigation but it is not viewed as substitutes for either arbitration or litigation.

Mediation

Mediation is a "...private, informal process in which disputants are assisted by one or more neutral third parties in their efforts toward settlement" (Stipanovich, 1996; citing Coulson, 1981). One or more impartial third-party mediators, or neutrals, facilitate mediation. Mediation is a structured, formal process, similar to arbitration, but not as expensive and time consuming as the arbitration process. A mediator cannot impose a settlement on the parties but the parties may choose, with the help of the mediator, to settle. It has been said that a successful mediation is one where both parties walk away equally unhappy. Mediation requires compromise between parties. When compromise doesn't happen, the mediation fails. Neither party is bound by the results.

Many contractors have had positive experiences with mediation. It is an effective dispute resolution technique. History has shown that mediated settlements are even more acceptable to the disputing parties when their mediator has had experience in construction law either as a judge, litigator or project principal.

Contract documents will frequently stipulate mediation to be administered by the American Arbitration Association in accordance with its Construction Industry Mediation Procedures (American Arbitration Association [AAA], 2009). The Association provides referrals to experienced mediators and their rules are posted on line at www.adr.org.

Mediators approach their work in different ways and some will alter their approach to the circumstances of the case and the dispositions of the disputing parties. Sometimes the mediator will be the conciliator, finding areas of agreement between the parties and encouraging them to settle. At other times the mediator may facilitate the parties into settling by explaining the time-consuming and costly consequences of not settling. The mediator may point out the strength and weaknesses of each party's case. The mediator may separate the disputing parties in different rooms and shuttle messages between them. It can be useful for the mediator to point out that if they do not reach a mediated settlement that will have to go to arbitration. Unlike a mediator, the arbitrator might pick one side or the other in a winner-take-all manner. This sometimes gets the mediating parties to soften their position.

Arbitration

Arbitration is a contractual choice. A written agreement to arbitrate must exist before the dispute occurs. Arbitration is conducted at a formal hearing in front of a professional arbitrator. It is the preferred method of binding dispute resolution method in the U.S. construction industry. Like mediation, a neutral party facilitates arbitration. Unlike mediation, the arbitrator, after hearing both sides of the issues, will impose a settlement upon the parties. The parties are legally bound to comply with the arbitrator's decision. The Federal Arbitration Act provides that choices to arbitrate are valid, legal, and enforceable (1925). The courts will enforce arbitrated dispute resolutions as if they were decisions reached by the court.

Demands for arbitration must be made in writing and delivered to the other party to the contract. The arbitrator may conduct a pre-hearing to ferret out all of the issues. In the arbitration hearing, the arbitrator hears both sides and makes a decision. Arbitration does not require compromise. The arbitrator makes a decision and that decision is binding on both parties. Some arbitrators will award a winner-take-all judgment to both parties. Others will split-the-difference and allow both parties to walk away with something.

Once decided, the parties are barred from litigating the same dispute in court. But either party can go to court after arbitration to obtain confirmation. Confirmation empowers the court to enforce the arbitrator's decision. A party can also go to court to vacate the arbitrator's decision. These challenges are rarely successful, however. Court's will almost always uphold an arbitrator's decision.

Subcontractors, sub-subcontractors, fabricators, suppliers, consultants and other project participants often become embroiled in the same disputes. This can lead to multiple arbitrations of the same disputes, with different parties, sometimes with contradictory results. Legal processes called <u>consolidation</u> or <u>joinder</u> enable the different parties to settle their issues at one time in a single arbitration. Consolidation or joinder is very desirable because it creates complementary settlements, reduces costs, and makes more efficient use of time and resources.

Although the cost of complex arbitration approaches the cost of a jury trial, most arbitration is not nearly as expensive or as time consuming as litigation. Moreover, arbitration is not bound by the complex rules of evidence that characterize litigation. The arbitrator need not even follow rules of law. The arbitrator may make decisions on elemental fairness as opposed to complex legal theories. Those are all good reasons to choose arbitration but the key benefit of arbitration may be that it is much faster than litigation.

The major weakness of arbitration is that arbitration decisions can rarely be appealed. Once the arbitrator makes a decision, it is final.

Selection of a competent arbitrator is very important. Simple disputes will involve a single arbitrator and a one-day hearing. Complex disputes will involve a panel of arbitrators and multiple hearings. The industry has a preference for arbitrators who are neutral experts on construction contract issues. Contract documents will frequently stipulate arbitration to be administered by the American Arbitration Association in accordance with its Construction Industry Arbitration Rules (AAA, 2009). The Association provides referrals to experienced arbitrators and their rules are posted on line at www.adr.org.

The arbitration rules of the International Chamber of Commerce (ICC) govern international construction disputes.

Litigation

The traditional method of resolving a dispute is through **litigation**, which is to say a lawsuit in whatever court has legal jurisdiction over the matter in dispute. Litigation is an expensive and complex process. Hundreds of different parties may participate in a single construction project: the owner, designer, subconsultants, construction manager, prime contractor, subcontractors, suppliers, sureties, insurers, and lenders. There will be layers of interdependencies between those parties. Claims will invariably touch many of those parties. Multiple parties will be drawn into the initial litigation through counterclaims, a claim asserted to oppose another party's claim, and joinder, the uniting of different claims into a single legal process. Because of counterclaims and joinder, the number of litigants grows and those litigants may find themselves unable to withdraw from a lawsuit, once it has begun.

Costly, lengthy and resource-consuming legal processes must be conducted before a trial can begin. A legal process known as **discovery** may be the most significant of the legal processes because it consumes so much of a contractor's time and attention. Discovery engages legal devices such as interrogatories, depositions, motions for entry upon land, and production of documents and things.

Interrogatories are written questions, prepared with the assistance of counsel, demanding written replies, also prepared with the assistance of counsel. Interrogatories are ordinarily exchanged between **claimant**, or **plaintiff**, the party whom is asserting the claim, and the **defendant**, the party who is opposing the claim.

A **deposition** is a record of live testimony given by a witness in advance of the trial in the presence of attorney representing plaintiff and defendant and a court reporter who creates a written record. Testimony in a deposition is given under oath. During a deposition, attorneys from both sides take their turns probing a witness to establish knowledge of the facts. Expert witnesses might be deposed for their knowledge and to explain complex issues such as critical path scheduling, cost accounting, material sciences, or construction management practices. Trade witnesses might be deposed for their intimate knowledge of jobsite safety practices, coordination and communication issues. Depositions usually have a disconcerting and confrontational tone.

A **motion for entry upon land** is a demand to make the jobsite, a warehouse, a fabrication shop or any other physical location available to the opposing party for a visual inspection. Because you can see, touch and smell the evidence, a visual inspection provides much more clarity than can be obtained through photographs or written documentation alone.

A **request for production of documents and things** is a demand for project records, memoranda, daily logs, computer files, photographs, and other such things. Typically, a request for production of documents and things accompanies an interrogatory. Documents and other physical evidence are needed to substantiate the replies made in response to an interrogatory.

During the trial, the attorneys will weave together a case built of legal theories and the evidence that they had gathered during discovery. That evidence will be explained and authenticated by live testimony. Attorneys will call witnesses, question their witnesses, and then the attorney for the opposition will question that witness, an activity known as cross-examination. The plaintiff has the burden of proof and the defendant need only rebut the plaintiff's proof. The standard of proof in a civil trial is usually preponderance of evidence, a much lower standard than that of a criminal trial. Both types of trials, criminal and civil, are adversarial by nature.

If, at the outset, both parties had waived their rights to a jury trial, the judge will hear the evidence, make factual determination, and make a ruling pursuant to law. This is called a bench trial. The reasons that the parties would choose a bench trial is the perception that a particular construction claim may be too complex for a jury of laypeople, hometown juries might be prejudiced or sympathetic, and that juries are notoriously unpredictable. The reasons that the parties would choose a jury trial may be that judges are more likely to accept disputed evidence and that a particular judge may not have experience with construction law. If there are no factual matters at issue with the claim, a judge may order a bench trial without the need for either party's consent.

The jury in a jury trial or the judge in a bench trial will arrive at a decision. Damages will be awarded to the winning party or specific performance will be ordered of the losing party and the court will use its authority to cause enforcement. Decisions are subject to appeal.

Arbitration vs. Litigation

Litigation is the default binding dispute resolution method in A201 (AIA, 2007b). Arbitration is the usual alternative binding dispute resolution method. Adequate contemplation must be given to the choice of arbitration or litigation as a binding dispute resolution method. There are significant differences between them, as follows and as summarized in Fig. 10.1:

- A lawyer must file a lawsuit with the court. Arbitration starts with the filing of a claim. Anybody can file a claim; it need not be a lawyer. Lawyers are required for litigation. Lawyers are not required for arbitration. That said; it is wise to have legal representation during arbitration.
- Filing fees for lawsuits are modest. The fees for starting arbitration are much higher. The courtroom and the services of the judges and clerks are provided free to the public. Arbitrators and their hearing rooms have to be paid for by somebody.
- Disputes that involve many parties are more economically and efficiently handled by one large litigation or arbitration, rather than as separate lawsuits or arbitration hearings. Multi-parties legal processes are created by consolidation or joinder. Consolidation or joinder is easy to accomplish in a lawsuit but it may not be so easy to accomplish in arbitration.
- Discovery is a process of gathering evidence. Discovery is mandatory in litigation. Until recently, however, one could not compel discovery in arbitration. Discovery is very time consuming and expensive. Worse, lawyers will use it to cause delays when it improves their bargaining position.
- Judges are experienced at conducting hearings. Arbitrators generally are less experienced at conducting hearings. Arbitration hearings, however, proceed much faster because they are not burdened with the restrictive rules of evidence found in judicial hearings.

- Legal hearings are going to be held where the lawsuit was filled and the court has jurisdiction over the dispute. The location of an arbitration hearing depends on the wording in the contract documents. Savvy owners will pick the most inconvenient arbitration location for the contractor.
- Arbitration hearings are private. Litigation is a public process. Sometimes the parties do not want the details of their disputes and their settlements published.
- Arbitrators generally are more experienced in construction matters than are judges.
- A party with a claim based on the language of the construction contract is more likely to succeed in litigation than in arbitration.
- Finally, some arbitrators seek to give both parties something rather than to decide all-or-nothing in one party's favor based on the merits of their claim.

Advantage	Arbitration	Litigation
No lawyer needed	✓	
Low filing fees and hearing costs		✓
Ease of consolidation and joinder		✓
Mandatory discovery		✓
High discovery cost avoided	✓	
Hearing experience		✓
Construction experience	✓	
Faster hearing	✓	
Location of hearing		✓
Private hearing	✓	
Decisions bases on contract language		✓
Everybody gets something	✓	

Figure 10.1 Arbitration vs. Litigation

All things considered, neither arbitration nor litigation has a clear advantage. Dissatisfaction with both dispute resolution methods has motivated the construction industry to experiment with other alternative dispute resolution methods.

CLAIMS & DISPUTE RESOLUTION

Federal Contract Disputes

The Contract Disputes Act [CDA] of 1978 governs dispute processes for construction contracts with federal government agencies (Contract Disputes Act [CDA], 1978). CDA claims are submitted to a contracting officer. The contracting officer makes the initial decision on the claim. If either the federal government agency or the contractor is unhappy with the initial decision, it may be appealed to either the Court of Federal Claims or to a Board of Contract Appeals, at the contractor's option (See Fig. 10.2).

A Board of Contract Appeals is a unique facet of federal agency contracting. A Board of Contract Appeals is composed of full-time professional attorneys at law. In their capacity as hearing officers and administrative judges they are vested with authority to administer oaths, issue subpoenas, hold hearings, examine witnesses, receive evidence, hear arguments, and write legal opinions for the purpose of deciding appeals of contract disputes with federal agencies.

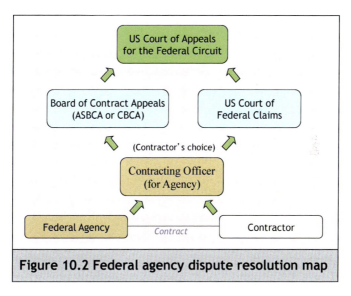

Figure 10.2 Federal agency dispute resolution map

Every federal agency used to have its own Board of Contract Appeals but they have more recently consolidated their operations. Since 2007, there are only two principal boards. The Armed Services Board of Contract Appeals [ASBCA] decides disputes for the Department of Defense, and the Departments of the Army, Navy, and Air Force. The Civilian Board of Contract Appeals [CBCA] is an independent tribunal established within the Government Services Administration [GSA] to serve every other federal agency except for the National Aeronautics and Space Administration, the United States Postal Service, the Postal Rate Commission and the Tennessee Valley Authority.

Contractors enjoy a right of direct access that empowers them to decide where appeals will be filed. Appealing to the court, however, is generally not a contractor's first choice. Benefits of appealing to a Board of Contract Appeals include: boards have demonstrated faster resolution of disputes than have courts; hearing officers and administrative judges generally have more knowledge and experience with public construction contracting and the provision of the FAR; and lower costs. Over the years, an enormous amount of case law has ascended from the various boards of contract appeals. Federal agency tribunals have shaped the contours of construction law jurisprudence. Both public and private sectors of the construction industry have benefited from this.

PART II - CONSTRUCTION RISKS

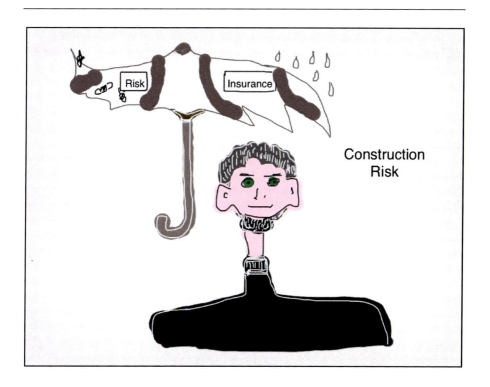

Chapter 11   BIDDING ON PUBLIC PROJECTS

11.1   Introduction

Through their own direct purchases, by providing grants and other funding to enable direct purchasing by other state and local governments and their agencies, or through its powerful and pervasive influence, our federal government and its agencies exert the single largest impact on design and construction activity in our industry (Keleher, 2008, p. 1). Yet most of the design and construction services needed by federal, state and local governments and their agencies, collectively known as **public agencies**, require capabilities that cannot be found within those agencies. They are purchased, instead, under contracts with private entities. Construction contracts between public agencies and private construction companies are known as **public construction contracts**.

Written public construction contracts cover the same ground as private construction contracts: the responsibilities of the owner, designer, contractor and subcontractors; payments and completion; concealed or unknown conditions; changes; time; interpretation of contract documents; suspension and termination; uncovering and correction of work; claims and dispute resolution; insurance; and surety bonds. And just as in the private sector, the parties to public construction contracts have implied duties and rights such as the duty to coordinate and cooperate, workmanlike performance, warranty of fitness, and good faith and fair dealing.

On the surface they look the same. The parties to both private and public construction contracts have express and implied duties and rights. Beneath that surface, however, there are layers of statutes and wide-ranging regulations by a multitude of public agencies that combine to shape public construction contracting processes. Contracting officers of public agencies must strictly adhere to highly formalized contracting processes. They not only risk their livelihood for violations but they expose themselves to criminal prosecution. Federal agencies do not cut corners. Nor will prudent contractors.

A public agency is a big dog. How do you treat a big dog? You give it whatever it wants. Nevertheless, the power of a public agency becomes restrained somewhat when they become a party to a public construction contract. Public agencies have equal status with private entities in judicial proceedings arising from public construction contracts (*McQuagge v. United States*, 1961). A public construction contract gives a public agency neither greater rights nor lesser duties than a common citizen.

Among the most important federal statutes is the Federal Acquisition Regulation (FAR, 1984). The FAR is not in itself a standard form contract although the FAR does contain standard contract clauses. Many of these standard contract clauses have found their way into private construction contracts. It is not unusual for a federal building construction project to use an AIA standard form contract including the A201 general conditions (AIA, 2007b). Supplementary conditions are normally used to incorporate FAR contract clauses into A201. Likewise, EJCDC C-700 might be used for engineered infrastructure projects along with appropriate supplementary conditions to incorporate elements of the FAR (EJCDC, 2007). Every state has its own department of transportation. State DOTs generally publish their own standard contract forms, largely modeled on principles expressed in the FAR.

The FAR also establishes the rules that federal agencies must follow during bidding, award, and administration of federal construction contracts. These rules have been adopted by many state and local governments and their agencies. Contractors first encounter these rules during the bid process for public construction contracts.

Another important piece of federal legislation is the Contract Disputes Act (the CDA) (Contract Disputes Act, 1978). The CDA and regulations that flow from it set out the rules for settling public contract claims and disputes. Boards of contract appeals established by public agencies routinely handle public construction contract claims and disputes. Decisions published by these agencies constitute the single largest body of law in the area of construction disputes (Keleher, 2008, p. 4).

The primary focus of this chapter is on the processes that are associated with bidding on public projects. A secondary focus of this chapter is on the anomalies that arise from bidding on public projects that lead to claims and disputes.

11.2 Bidding processes

Each public agency's bidding processes are managed by a **contracting officer** [CO] or officers in their employment who are vested with the authority to solicit for bids, evaluate bids, make contract awards and to administer those contracts on behalf of the public agency.

The FAR mandates that a CO must purchase construction services at *fair and reasonable* prices, from responsible contractors (FAR, § 15.402(a), 1984). Internal cost estimates, prepared by the CO or by the public agency's staff help to establish the reasonableness of any bid. COs have considerable discretion though and may or may not reject any or all bids based on their proximity to an internal estimate.

Most public agencies will invite potential bidders to a **pre-bid conference** where they can answer questions posed by the bidders and communicate additional information relevant to the bid. A solicitation may require written questions to be submitted in advance to the CO or other point of contact named in the solicitation. Whatever the rules, it behooves bidders to strictly adhere to them or risk having their bid rejected if they do not.

If a question, properly raised, is not answered or the answer is not meaningful, a contractor should consider filing a bid protest. But any information that is communicated during the pre-bid conference can be relied upon, provided that it is clearly communicated and later distilled into writing and issued to the bidders. Contractors should not rely upon any verbal communications.

Responsive Bidder

A **responsive** bid is one that complies with all material requirements of a public agency's solicitation. Failure to comply with each and every material requirement renders a bid **nonresponsive**. Nonresponsive bids cannot be considered. If a deviation making a bid nonresponsive is not discovered until after the award the effected contract is voidable or alternatively, the compensation under that contract may be reduced.

To depict a bid as nonresponsive a deviation must be "material." A **material deviation** is an irregularity that gives a bidder a substantial competitive advantage; prevents other bidders from competing equally; affects price quantity, quality or delivery; or otherwise prejudices the other bidders (FAR, § 14.404-2(d), 1984). Material deviations are found in the failure to state a price, an indefinite price, pricing established at a future date or in qualifications such as claiming the right to a price adjustment if certain conditions occur, the right to future price adjustments in response to increased costs, or any assertion that limits the rights or modifies the duties of the public agency.

According to Kelleher, Abernathy & Bell, "...the concept of bid "responsiveness" is used to guard against the low bidder's having the opportunity, after bids are opened and all prices are revealed, to accept or reject an award based on some contingency that the bidder created itself and that only applies to, and works to the advantage of, that bidder" (2008, p. 68). The phrase "two bites at the apple" is commonly used to describe any bidder who gains a second opportunity, after the bid opening, to seize an advantage.

A bidder should not insert its own clarifications, qualifications or explanations into its bid. These can be deemed material deviations. Substitutions identified in a bidder's proposal can be a source of material deviation. A **substitution** is any equipment, material, article or process that differs from what is prescribed in the specifications. FAR § 52.236-5 explains that the trade names, make, or catalog numbers referenced in specifications, "...shall be regarded as establishing a standard of quality and shall not be construed as limiting competition (FAR, 1984). The Contractor may, at its option, use any equipment, material, article, or process that, in the judgment of the Contracting Officer, is equal to that named in the specifications, unless otherwise specifically provided in this contract." This clause opens the door to substitution. But the caveat in the § 52.236-5 "or equal" clause is that when a bidder exercises his option to substitute, its substitute may or may not be deemed equal by the CO. Substitutions identified in a contractor's bid may render a bid nonresponsive (FAR, 1984).

The equipment, material, articles and processes specified in the construction documents should be bid exactly as specified. This "exactly as specified" bid is known as the "base bid." Sometimes, the bid instructions will also solicit pricing for alternates. **Alternates** consist of different equipment, material, articles or processes than those prescribed in the construction documents. Each alternate adds to or deducts from a bidding contractor's base bid.

Another type of alternate applies to large projects. Large projects might be broken down into phases, each constituting a small subset of the entire project, where phase 1 constitutes the base bid and subsequent phases are alternates. This technique provides flexibility to public agencies that may be unsure of future funding allocations.

A contractor whose base bid is low may no longer be the low bidder after alternates are added or deducted from its base bid. In the interest of fairness, public agencies will state in their bid instructions their preferences for alternates and assert them as determining factor upon which award will be made. Accordingly, failure to provide a base bid and to price each and every alternate may render a bid nonresponsive.

An **addenda** is any change in the bid solicitation that is issued before the bids are submitted. Bidders must list each and every addenda on their bid forms and provide written acknowledgment of their receipt and incorporation into their bid amount. Failure to acknowledge each and every addenda may render a bid nonresponsive.

Another type of material deviation can arise with front-loaded bids. Front-loaded bids are common with unit-price contracts. A **front-loaded bid** is one that utilizes any pricing technique that modifies unit-prices, or the schedule of values, to increase the amounts paid to a contractor at the beginning of a project while reducing payments at the end of a project. There is no net gain to a contractor as a consequence of a front-loaded bid. Nevertheless, increased early cash flow lowers a contractors' financing costs. This may be deemed an unfair advantage with respect to other bidders and therefore render a bid nonresponsive.

In some states a CO has discretion to overlook minor deviations in the bid. Minor deviations must be shown to be so trifling or unimportant that no competitive advantage could possibly be gained. But what a bidder considers to be minor may be material to the CO and courts tend to defer to a COs judgment. It behooves bidders for public construction contracts to meticulously scrub every potential deviation out of their bid language.

Responsible Bidder

A **responsible** bidder is one who has the necessary character, experience and capability to perform the work of the contract. To demonstrate responsibility, FAR § 9.104-2 requires a bidder to:

(1) Have adequate financial resources to perform the contract, or the ability to obtain them;
(2) Be able to comply with the required or proposed delivery or performance schedule, taking into account all existing commercial and governmental business commitments;
(3) Have a satisfactory performance record;
(4) Have a satisfactory record of integrity and business ethics (emphasis added);
(5) Have the necessary organization, experience, accounting and operational controls, and technical skills, or the ability to obtain them;

(6) Have the necessary production, construction, and technical equipment and facilities, or be able to obtain them;
(7) Be otherwise qualified and eligible for award under applicable laws and regulations (FAR, 1984).

Bidders must supply sufficient current and complete information to enable the public agency to make a responsibility determination. This determination coalesces into two distinctive elements: 1) the future ability of the contractor to perform the work of the contract; and 2) the bidder's past performance record with public construction contracts.

In the evaluation of future ability, the focus will not be on the bidders capabilities at the time of bidding but on the capabilities the bidder has or can reasonably acquire before the work of the contract commences. A bidder may need to demonstrate that capable employees and other resources can be acquired, either by direct hire or by subcontracting.

Since 1994, the federal government has been keeping records of contractor **performance evaluations** on all contracts for more than $100,000. These past performance evaluations are major factors in the responsibility determinations on current projects. Performance evaluations are also important management tools because a contractor who is motivated to earn a high performance evaluation tends to create higher levels of customer satisfaction as a bi-product of superior performance.

Criminal behavior or other wrongful behavior by a contractor may result in its **suspension** or **debarment** by a public agency. Debarment prohibits a contractor from bidding on or performing public construction contracts for a specified period of time, but not to exceed three years, the maximum term for a debarment. Suspensions are temporary disqualifications that are used while the agency investigates or while a legal proceeding takes place.

The CO does not make suspension and debarment decisions. The head of a public agency makes those decisions. Grounds for debarment include actions and inducements such as: commission of fraud; a criminal offense showing a lack of business integrity; a serious violation of contract terms; other behaviors showing a lack of present responsibility; knowingly doing business with an ineligible person; failing to pay debts to any federal agency (except the Internal Revenue Service); willful violation of a statute or regulation applicable to a public agreement; and violation of a voluntary exclusion agreement or a debarment or suspensions settlement agreement.

BIDDING ON PUBLIC PROJECTS

Competitive Sealed Bidding

Competitive sealed bidding is the usual way that public agencies solicit bids for construction. The process begins when a public agency issues an **invitation for bids [IFB].** The IFB consists of written bid instructions; directions for obtaining or gaining access to a complete set of construction documents (CDs); bid forms; the bid due date; and the date, time and place where the bids will be opened. Usually, bids must be delivered at a specified date, time and place in a sealed envelope. At the bid opening, the envelopes are opened and the bids are announced to the public in attendance. The contract award goes to the low responsive and responsible bidder.

The IFB assures that the same information is supplied to every bidder, it encourages uniformity of bid content and format, and it reduces the time and effort required to review bids. A typical IFB will provide the following information:

- Instructions on how to prepare a bid, what document format to use, and what forms must be completed;
- Submittal instructions including packaging, the address that the bid package must be delivered to, the date and time of submittal, the number of copies, and other particulars.
- Rules for receiving and acknowledging **addenda**, defined as changes in the construction documents or bid instructions, distributed to bidders prior to the bid due date.
- Requirements for demonstrating the qualifications of the bidder, such as the bidder's financial, management, and administrative capacity and past performance record;
- If **bid security** is required, an indication of whether that security must be in the form of a surety bond or a certified bank check, along with the amount of security, provisions for forfeiture, provisions for return of bid security to unsuccessful bidders and other such details;
- Conditions for allowing bid withdrawals, procedures for withdrawing bids, penalties, if any, and time limits thereto;
- The date, time and place for opening bids and a declaration of whether the bidders information will be made public or remain private;
- Conditions for rejection of bids, whether or not reasons and justification will be provided for rejecting bids, and procedures for resolving bid discrepancies;

- Time limitations including the period of validity of bids and the project timeframe in working days or calendar days; and
- Special rules, regulations and requirements unique to the project, such as set-asides, working hours, purchasing provisions, construction phases, and special activity sequences.

A designer, working on behalf of the public agency, usually prepares the IFB. Increasingly, construction managers fulfill this function. Generally, the party who prepares the IFB will also be responsible for its distribution, interpretation, corrections, collection, review, and evaluation. Regardless of who prepares it, the IFB must describe the public agency's requirements clearly, completely and accurately. This is especially important with competitive sealed bidding because the public agency will not have the opportunity to evaluate the technical merits of any bid.

Competitive sealed bidding provides a public agency with the benefit of competitive pricing and it can also ensure the largest pool of prospective bidders. To do so requires distribution of the IFB to known contractors, posting of solicitations in public places and advertising through reporting services. A secondary benefit of competitive sealed bidding is that it proves difficult for patronage, nepotism, collusion or fraud to gain traction when public construction contracts are awarded through competitive sealed bids responding to widely publicized solicitations.

Best Value Procurement

An increasing number of public construction contracts are being awarded through a bidding process known as **best value procurement**. Best value procurement enables a public agency to define their best interests through a trade-off process between cost or price and non-cost factors. The enabling language in FAR § 15.101-1, is reprinted below:

(a) A tradeoff process is appropriate when it may be in the best interest of the [public agency] to consider award to other than the lowest priced offeror or other than the highest technically rated offeror.
(b) When using a tradeoff process, the following apply:
    (1) All evaluation factors and significant subfactors that will affect contract award and their relative importance shall be clearly stated in the solicitation; and
    (2) The solicitation shall state whether all evaluation factors other than cost or price, when combined, are significant-

ly more important than, approximately equal to, or significantly less important than cost or price.

(c) This process permits tradeoffs among cost or price and non-cost factors and allows [a public agency] to accept other than the lowest priced proposal. The perceived benefits of the higher priced proposal shall merit the additional cost, and the rationale for tradeoffs must be documented in the file in accordance with FAR §15.406 (FAR, 1984).

The solicitation in best value procurement is a form of **negotiated procurement**. In negotiated procurement, an agency issues a **request for proposals [RFP]**. An RFP is similar to an IFB except that proposals responding to an RFP are not handled the way that competitive bids are handled with respect to an IFB. After receiving proposals responding to an RFP the public agency enters into "discussions," otherwise called negotiations, with the contractor or contractors on a so-called "short-list" of proposals that fall within a predetermined "competitive range." Proposals that fall outside of the competitive range are rejected. Following discussions, short-listed contractors are given the opportunity to submit final proposal revisions, commonly known as "best and final" offers. The public agency is required to fix a "common cutoff date" for receipt of final proposal revisions. And finally, after its review of all final proposal revisions it makes its award to the contractor whose proposal meets the public agency's best interest, the criteria for which are defined in the RFP.

Best value procurement is negotiated procurement that may or may not involve a short-list followed by discussions. The public agency may choose to award without discussions provided that it so informs bidders of that possibility in its RFP. In the event that an award without discussion is contemplated, the public agency may also seek "clarification" through written communication or a face-to-face meeting with the supposed winning contractor, prior to award. In the event that the public agency chooses to short-list and discuss, the type of topics that can be discussed and the limits of those discussions are set out in FAR §§ 15.306(d) and (e). Other types of communications that are authorized between any short-listed offeror and the public agency are set out in FAR §§ 15.306(b) (FAR, 1984).

Project Delivery Strategies

Design-Bid-Build

Design-bid-build [DBB] is the traditional project delivery strategy of public construction contracting. DBB is a sequential process: first the project's design must be completed, next there is a sealed competitive bid, and finally the contractor starts building, hence the moniker design-bid-build. DBB is an established process, well understood by all involved parties.

Although a sealed competitive bid with a firm-fixed price is the norm, DBB can also be configured as firm-fixed price with economic price adjustments or as a unit-price contract. Public agencies will choose an appropriate contract pricing method for the particular circumstances of the work.

DBB has its shortcomings. One shortcoming arises from the design process. State-of-the-art design solutions arise from two entrées: 1) access to the proprietary designs for equipment, materials and systems provided by industrial manufacturers; and 2) the participation of the licensed specialty subcontractors who install and warrant those manufacturers' products. Designers engaged in the traditional DBB process lack unfettered access to both entrées. Thus, DBB tends to generate outmoded design solutions with lower levels of quality, ease of assembly and economy. Outmoded design becomes increasingly more problematic with complex criteria definitions.

Another shortcoming is time. DBB is sequential. Significant opportunity costs are lost while construction waits for the design process to go to 100% completion before the bid process starts and actual construction can begin. Worse, should all of the bids come in over budget, something that occurs all too often, the agency must go back to the designers to have their design revised and made less expensive, and then start all over again with bidding. DBB can be frustratingly time consuming…and expensive.

Yet another shortcoming arises from the construction process itself. It can be adversarial to excess. This is particularly the case whenever the construction documents are inadequately developed or there is an abundance of performance specifications with indefinite performance criteria. Contractors laboring under cutthroat, competitive pricing will seize every opportunity for relief through change orders, precipitating highly adversarial project relationships.

All things considered, DBB remains a satisfactory project delivery strategy for routine projects that possess fully detailed construction drawings and complete prescriptive specifications, particularly when time is not the primary concern.

Design-Build

Design-build [DB] contracts are authorized under FAR § 36.3 (FAR, 1984). A DB contractor, or design-builder, will take responsibility for both design and construction. But first, the public agency must describe its intended scope of design and construction work and otherwise define its needs. This is accomplished with **criteria documents**. Typically, the public agency, under separate contract with a design firm, develops criteria documents. Once that is accomplished, a two-part bid process begins.

During part one of the process, the public agency solicits prequalification information from interested design builders. Responses are used to build a short-list of three to five of the most qualified firms. Firms are not to respond with prices or technical solutions. The public agency must build their short-list based solely upon the prequalification responses, such as evidence of prior experience performing the type of design and construction work that will be solicited in part two. Ordinarily, the part one prequalification solicitation, commonly known as a **request for qualifications** [RFQ], will also define the scope of work of part two, identify both part one and part two evaluation factors, and assert the number of candidates that will be selected for the short-list.

During part two the short-listed candidates are invited to submit two separate competitive proposals in response to a **request for proposals [RFP]**. One of those proposals is a technical proposal, describing the design builder's technical response to the criteria documents. The other proposal is a cost or price proposal. The public agency then picks the design builder whose proposals meet their best interests.

Cost-plus-fixed-fee pricing with or without incentives is the norm for DB contracts but they can also be configured with firm-fixed-pricing, with or without economic price adjustment.

DB has its shortcoming. One arises from a hand-off of design control. The public agency must rely on their criteria documents to describe their needs with sufficient clarity because apart from the criteria documents the public agency will have little, if any, influence over the design. The design builder will design to the criteria documents and respond to the pricing strategy thereof, not necessarily to direction by the public agency.

Another shortcoming arises from reactions to the first shortcoming. A public agency wishing to retain more control over design may actually do some of the design and include that in its RFP. Such post-criteria design work is known as **bridging**. The designer who created the criteria documents is ordinarily retained for bridging. Bridging documents can include conceptual design documents, schematic design documents, design development documents and even some of the construction documents and specifications. Up to 70% of a design have been completed with bridging. But the more bridging, the more a DB project resembles DBB and all of the shortcomings that go with that.

Yet another shortcoming arises from organizational management. DB solicitations from public agencies have a penchant for one-of-a-kind projects filled with special requirements. The firms that respond tend to be ad hoc teams, assembled from many different firms, each with different specializations, who may have or may not have ever worked together before. DB is a collaborative process between teams with very different interests. For collaboration to prevail over self-interest, working relationships must be founded on trust. Trust, however, is not a commodity; it is an interpersonal phenomenon (Child, Faulkner and Tallman, 2005). Whenever dysfunction sets in amongst DB teammates, inferior work products result.

All things considered, DB can produce extraordinary results. It has the capacity to bring state-of-the art design solutions to reality. Early involvement of the contractor avoids rushing the subcontractor bids, enabling a thorough vetting of subcontractors with lower costs, timely performance, and higher quality as its consequence. It is a particularly good project delivery strategy for **fast tracking**. Fast tracked projects overlap design and construction activities in order to significantly reduce time schedules from the commencement of design to final completion. There have been many DB projects by public agencies that have demonstrated significant cost savings, faster delivery schedules and improved quality.

Construction Management

Construction management [CM] is yet another project delivery strategy employed by public agencies. This strategy superimposes a CM layer over what would otherwise be a DBB project. Among other things, a construction manager provides **preconstruction services.** Preconstruction services are services related to design and planning activities such as: an evaluation of the project's program, schedule and budget; preliminary cost estimates; phasing plans; identification of long-lead items for procurement; preparation of bidder's lists; and bid evaluations.

In the U.S., architects, contractors, engineers or other consultants, either as individuals or as firms, may identify themselves as construction managers. Most construction manager are not licensed as such in the U.S. In England and many commonwealth countries though, **chartered surveyors,** whom are licensed and registered professionals, provide CM services. Within the international building community, chartered surveyors hold a status much like that enjoyed by registered architects or engineers in the U.S.

The purpose of the CM project delivery strategy is to provide the public agency with independent, third-party consultation during design and construction. Early applications of the CM project delivery strategy had the construction manager providing preconstruction services during the design phase and administrative services, similar to what an architect or engineer would ordinarily provide in DBB, during the construction phase. This came to be known as **CM-agency [CMa]**, alluding to a contractual agreement that installed the construction manager as an agent of the public agency, as principal. This "agency agreement" established a form of **fiduciary** relationship. A fiduciary is an agent with superior knowledge and experience who is bound by a duty of trust and confidence to do what is in the best interest of the principal.

Although the role of the construction manager in CMa was to steer the ship through the hazards of design and construction - the major hazard being failure to complete on time and under budget - the construction manager stopped short of guaranteeing budget or schedule. The public agency retains the risk of cost and schedule in the CMa project delivery strategy. Architects or engineers stepped easily into this role because it mimicked their historic role in administering construction contracts, a common practice with the DBB project delivery strategy.

In time, public agencies began to seek price and performance guarantees from CM firms. General contractors operating as construction managers proved willing, under certain circumstances, to provide these guarantees. A new project delivery strategy evolved, known as **CM-at-risk [CMAR]**. This strategy distinctively differs from CMa in that cost and schedule risk is transferred from the public agency to the construction manager.

For CMAR, a public agency prequalifies construction managers whom have capability to provide both preconstruction services and actual construction. The selected construction manager then provides preconstruction services during the design process using cost-plus-fixed fee pricing. At a point in time when the construction documents are developed to the point of describing the project in sufficient detail, the construction manager provides a price for construction. The pricing method is usually cost-plus-fixed fee with a ceiling price, or cap that may not be exceeded. If this ceiling price, referred to as a **guaranteed maximum price [GMP]**, exceeds the public agency's budget, they can either order a redesign to budget, at their expense, or stop the work and terminate all contracts. If the project does continue, a construction contract is executed with the construction manager, whose role comes to resemble the traditional role of a general contractor.

There is much debate over whether CMa or CMAR is the best project delivery strategy, but in states that allow their public agencies to use CMAR there are few CMa projects (Thomsen & Sanders, 2011, p. 183).

CMAR has its shortcomings. The CM layer adds additional costs. Proponents argue correctly that as a general contractor, the construction manager has access to industrial manufacturers and licensed specialty subcontractors and that this results in higher quality, ease of assembly and lower costs. These benefits more than compensate for the additional cost of the CM layer, they argue. But it is often difficult to demonstrate these benefits because CMAR projects tend to be unique projects that do not possess a good baseline for comparison. It is instructive to note that some construction managers will charge little or nothing for preconstruction services, in effect giving them away in exchange for nothing more than the inside track to a construction contract award.

Another shortcoming arises if the construction manager's fiduciary relationship with the public agency during the design process dissolves into self-interest, as the construction manager becomes the general contractor and assumes cost and schedule risk. The public agency may become burdened with additional oversight as it tries to administer the construction contract of its construction manager turned general contractor.

Yet another shortcoming relates to the same organizational management concerns that loom over DB project delivery. At least in DB projects, some designers and builders have a healthy desire to work together. But CMAR is usually conceived as a shotgun marriage between architect/engineer and construction manager/general contractor where the construction manager is brought in, given a mandate over design but no contractual authority thereto. Whenever dysfunction sets in between the design firm or firms and the construction manager or its partners, inferior work products result.

All things considered, CMAR can produce extraordinary results. It has the capacity to bring state-of-the art design solutions to reality. Early involvement of the construction manager avoids rushing the subcontractor bids, enabling a thorough vetting of subcontractors with lower costs, timely performance, and higher quality as its consequence. Public agencies do not risk being hamstrung by their own criteria documents and they enjoy more control over the design process. There have been many CMAR projects by public agencies that have demonstrated significant cost savings, improved quality and greater satisfaction by the public agency.

A handful of projects have recently been completed by the U.S. Army Corps of Engineers using an **integrated-design-bid-build [IDBB]** project delivery strategy. IDBB is a hybrid of different project delivery strategies. It ultimately has a firm-fixed-price construction contract, like a typical DBB project; it also has a target price and GMP that are developed early on, much like with DB; and the constructor and the Corps have an agency relationship, much like they do in CMa or CMAR. Implementation starts with a qualifications-based selection of a design firm. (The Federal Brooks Act prohibits public agencies from soliciting a design contract based on competitive pricing). Upon completion of schematic design (roughly 15% of design) they then issue an RFP for construction management with cost-plus-fixed-fee plus GMP pricing against an initial target cost. The winning construction manager comes on board and provides preconstruction services pursuant to an agency agreement with the Corps. Then, with 100% of design documents complete, a point deemed the "production point," the Corps and the construction manager negotiate a firm-fixed-price where the profit is adjusted based upon the relationship between the initial target price and the firm-fixed-price.

IDBB has its shortcomings. It is new and not widely understood. Some contractors have concluded that IDDB projects are too risky to bid, citing equivocality in the relationship between the initial target price and the firm-fixed price, and the lack of an escape clause in IDDB contracts (U.S. Army Corps of Engineers and North Atlantic Regional Business Center, 2008). To the Corps, IDDB presents the double-edge sword of increased Corps involvement in design - a benefit - while demanding significantly more resources from the Corps to facilitate their involvement - a detractor (2008). Nevertheless, the Corps has successfully applied this strategy on a handful of projects. Time will tell if it gains traction with other public agencies.

Public Private Partnership

Public private partnership [P3] is a unique project delivery strategy. A P3 is a package of contractual agreements between private entities and a public agency to finance, design, build, commission, operate and maintain. P3 has been applied to public buildings and to infrastructure projects involving transportation, utilities and water resources. It can involve both new construction and the renovation of existing facilities.

A P3 is made possible by a **project concession agreement** between a private entity and a public agency whereby the private entity commits to finance, design, build, commission, operate and maintain the project for a period of years. That private entity, enabled by various shareholders and project sponsors, forms a private consortium of banks or other private funding sources, designers, contractors and facilities operators. Revenues earned, taxes generated, or costs savings during operation provide compensation to the private consortium. The public agency retains ownership of the project throughout, while the private consortium assumes project risk in exchange for an income stream over a period of years. At the end of the period of years there is a hand back of facility operation and maintenance to the public agency, at which point the private consortium exits.

P3 has it shortcomings. As with design-build, there is a hand off of design control from the public agency to the private entity. But a public agency has even less influence over design in a P3 because the public entity must protect its own financial interests in operating and maintaining the facility. P3s are sometimes criticized for sidestepping public policy initiatives, such as small business set-asides, minority-owned business program, and the Historically Underutilized Business Zone (HUBZone) program. Nevertheless, P3s represent hope for public agencies: to build or renovate facilities despite severe fiscal restraints and to transfer risks to the private sector while avoiding the cost and schedule overruns that plague public procurement.

Job Order Contracting

Job order contracting began in the Air Force in the 1980s. It has since been adopted by many states. It is used primarily for small maintenance and repair projects. Its pricing method is unique. Job order contractors agree to do whatever the agency assigns to them over a set period of time while the agency agrees to pay for that work according to a pricing formula that gets established at the outset.

The way it works is the public agency publishes a price book of unit prices for every construction operation that could conceivably be incorporated into the projects that they are contemplating. This price book is sometimes specific to the particular agency but more often it is a published standard, such as RSMeans™ assembly prices. The job order contractor is contractually committed to a multiplier that covers its overhead, profit and any risk premium. Each job order is priced by multiplying the contractor's multiplier by the book price for each construction operation in the scope of work of the particular job order.

The solicitation for job order contracting is done with an RFQ. The bidding instructions provide example work scopes for the types of projects contemplated by the agency, the price book, and the minimum and maximum amounts that the total contract is valued at. Bidders submit their qualification and a multiplier. Public agencies evaluate potential contractors by their qualifications and then select either a qualified contractor with the lowest multiplier, or the contractor who is believed to deliver the best value considering qualification, interviews of key personnel, and its multiplier. A contract is then executed with the successful bidder that locks in its multiplier over a specified period of time.

Job order contracting has its shortcomings. For all but the simplest of projects, the actual cost to build will be something other than the sum total of it component parts. A contractor bidding a DBB project is able to adjust its price based upon judgment and experience. There is no opportunity to do that with a job order bid. A separate issue is that the job order contractor commits to do every job and any job that the agency orders during the term of its contract. One job might be large and routine, producing economies of scale, another might be small and complex, costly to execute. The job order contractor cannot select one or the other. It must do every job, not matter if it makes money or loses money on the job. Perhaps the most problematic issue is that the job order contractor may be subjected to dramatic swings in demand because it cannot control the size or frequency of the jobs that it is required to do.

Job order contracting is simple though. It is great for routine, small, and uncomplicated projects. It reduces the time and cost of procurement. And once the job order contractor is on board, work can start very quickly on any project.

e-Bidding

FAR §15 authorizes electronic bidding, or **e-bidding**, a term that denotes the submittal of bids or proposals by transmitting data or document files over the internet as opposed to conveying printed documents to a physical location (FAR, 1984). Although §15 deals with negotiated procurement, it enables e-bidding as a alternative to both competitive sealed bidding and negotiated procurement.

A public agency's usual justification for e-bidding is that it provides contractors with a beneficial alternative to conveying a bid in a sealed envelope. Contractors gain more time to fine-tune their proposals by eliminating the time they would otherwise waste to physically deliver their bid to the public agency. Public agencies benefit by receiving higher quality bids, with fewer errors or omissions and lower incidences of bid mistakes.

Prudent bidders will submit electronic bids well in advance of the published deadline and seek verification that their bid was received on time. They know - or they have discovered it the hard way - that digital transmission is not "instantaneous" transmission. Minutes and hours can elapse during data center traffic jams, with digital transmissions bottlenecked in the switches, routers and servers for virus-scanning, encryption, password authentication and other routines, before actually being received at the electronic "location" specified in the bid instructions, such as the CO's mailbox. The Comptroller General has ruled that electronic bids not "received" at the specified location on time are late, therefore nonresponsive, regardless of having been sent before the deadline and no matter that the contractor was not at fault (*Sea Box, Inc.*, 2002).

11.3   Bid Irregularities

Especially with competitive sealed bidding for firm-fixed-contracts, where subcontractor bids arrive at the very last minute and the prime contractor's bid is hastily prepared, bidders want to know that relief is possible for the mistakes that they will inevitably make.

Bid mistakes made solely by a contractor are called unilateral mistakes. Relief is available for certain types of unilateral mistakes as long as they are noticed before the bid is awarded. Relief is generally not available for unilateral mistakes that are noticed after the bid is awarded.

Unilateral Mistakes

In the construction context, a unilateral mistake is one that is made by a bidding contractor. Unilateral mistake invokes three different legal theories: clerical mistakes, mistakes in judgment, and unconscionability.

FAR § 14.407-2(a) defines **clerical mistakes**, as follows:
"(a)   ... Examples of apparent mistakes are --
   (1) Obvious misplacement of a decimal point;
   (2) Obviously incorrect discounts (for example, 1 percent 10 days, 2 percent 20 days, 5 percent 30 days);
   (3) Obvious reversal of the price f.o.b. destination and price f.o.b. origin; and
   (4) Obvious mistake in designation of unit..." (FAR, 1984)

Recognizing that clerical mistakes occur during the last minute rush to bid, the FAR provides some relief to bidders for their clerical mistakes.

Mistakes in judgment, in contrast, go to a contractor's basic competency. A contractor must suffer its own mistakes in judgment. The Court of Federal Claims explains that, "...mistakes in judgment on the part of [a bidding] contractor are not the types of mistakes that are compensable." (*Lakeshore*, 2013) So if a contractor applied the wrong mathematical formulas and thereby miscalculated the volume of concrete required, this mistake in judgment would be at the contractor's expense, but if the contractor, having calculated correctly, mistakenly transcribed an incorrect volume onto the bid forms, it would be considered a "clerical mistake" for which the contractor may seek relief.

The onus for noticing a unilateral mistake is not entirely on the contractor. The government must act whenever it notices or should have noticed an apparent mistake. **Apparent mistake** arises when a unilateral mistake was obvious on the face of the bid documents and, in the words of the Court of Federal Claims, "...the contracting officer knew or should have known of the contractor's...mistake at the time that the bid was accepted..." (*Lakeshore,* 2013). The FAR compels a CO to inform a bidder when a unilateral mistake is suspected and require the bidder to verify its intended bid. On failing to do so, the contractor may be entitled to relief. In the *Appeal of Pavco, Inc*. the Armed Services Board of Contract Appeals reasoned, "It would be unfair or unconscionable for the Government to take advantage of a contractor's mistake in bid if the Government knew or should have known of the mistake prior to contract award" (1980). The board's conclusion is sometimes referred to as the "snap-up" doctrine; alluding to a buyer who rushes to make an award in order to snap up an unearned benefit from an unwitting seller.

Mistakes Noticed Before Award

After the bids are opened the CO will examine them for mistakes. When the CO has reason to believe that a mistake may have been made, the CO will request verification of the bidder's intended bid. If the bidder replies that its bid is not in error, the CO will establish the bid as it was originally submitted. But if the bidder replies that its bid was in error, clear and convincing evidence must then be provided to establish both the existence of the mistake and the bid actually intended. Suitable evidence includes the bidder's file copy of the bid, the original worksheets and other data used in preparing the bid, subcontractors' and suppliers' quotations, if any, published price lists, and any other evidence that establishes the existence of the mistake, the manner in which it occurred, and the bid amount actually intended.

A bidder's clerical mistake may be apparent on the face of the bid documents. Bids infer clerical mistakes when they are so far out of line with the amounts of other bids, or with the amount estimated by the agency, or as determined by the contracting officer to be reasonable, or show other clear indications of error. The CO will correct a bidder's apparent clerical mistake, provided that the bidder confirms that its bid was indeed in error and provides clear and convincing evidence establishing the bid that was actually intended.

The bidder may be the first to allege a mistake. Upon notification, the CO must first determine if the bid was responsive. Allegations of mistakes may not be used to permit correction of a bid to make it responsive. Provided that the bid was responsive, the CO will advise the bidder to make a written request to either withdraw or correct the bid and to provide clear and convincing evidence to establish both the existence of the mistake and the bid actually intended. Upon being provided with evidence, the CO will evaluate it, but the agency head, not the CO, will make the determination in these cases.

If the agency head finds that the evidence provided was NOT clear and convincing as to the existence of a mistake or the bid actually intended, the determination may be that the bid can be neither withdrawn nor corrected. But if, "...(1) The evidence of a mistake is clear and convincing only as to the mistake but not as to the intended bid, or (2) The evidence reasonably supports the existence of a mistake but is not clear and convincing..." the determination may be to permit the bidder to withdraw its bid but not to correct it (FAR, § 14.407-3(b)(2), 1984).

When the bidder establishes clear and convincing evidence of a mistake and its intended bid, different determinations are possible. If the bidder had requested permission to correct the mistake, the agency head may allow it, unless that correction would result in displacing one or more lower bids. An exception is made when the error was a clerical mistake where, "...the existence of the mistake and the bid actually intended are ascertainable substantially from the invitation and the bid itself" (Federal Acquisition Regulation, § 14.407-3(a), 2013). Apparent clerical mistakes may be corrected even if that correction would result in displacing one or more lower bids. If the bidder had requested permission to withdraw the bid rather than correct it, the agency head may allow it, unless the bid, both as uncorrected and as corrected, is the lowest received. In that event the determination may be to correct the bid and not permit its withdrawal.

Mutual Mistake

There is no relief for a unilateral mistake once the contract award is made, unless the CO knew or should have known of a contractor's bid mistake at the time that the bid was accepted. There might be relief after award for a mutual mistake. Although the actual instances of mutual mistake are few and far between, if a contractor can demonstrate that a mutual mistake was made, its contract may be reformed or rescinded after award.

A **mutual mistake** arises in law under two different contexts: 1) when there had been agreement between parties but the written form of their agreement does not express what was actually intended by them; or 2) there had been no true agreement between the parties because of a misunderstanding of a material fact.

The *Ballard v. Loving* case provides a practical example of context #1 (1994). In this case, Ballard and Sons, a general contractor, had a contract to build phase one of the Loving High School. At the same time that the construction contract was signed, a deductive change order was signed to eliminate paving of certain parking areas. Neither party realized, at that time, that the contract price had already been reduced for the eliminated paving. The parties had unwittingly deducted the paving twice. The court ordered the contract reformed to compensate Ballard for its loss.

Students of construction law should understand that a contract reformation is not the same as a change order. A change order is an agreement between the parties that is supported by consideration, both parties enjoying benefit and suffering detriment. A reformation is different because consideration is irrelevant. In Ballard, when the court determined that the written form of the construction contract did not express what the parties actually intended, it rewrote the contract to Ballard's benefit. To reform a contract is to rewrite it as it was intended to be notwithstanding any new consideration..

The case of *Native Homes v Stamm* provides an example of context #2 (1998). In this case, Stamm, a homeowner entered into a contract with Native Homes to build a new residence on a clean site. Stamm had previously hired an excavation contractor to remove muck from the site and replaced with clean, compacted fill. After notice to proceed but before any actual construction the contractor discovered that the muck and fill work had never been performed. The Florida court allowed the homeowner to rescind the construction contract, finding that there was a mutual misunderstanding of a material fact.

Students of construction law should understand that rescission is not the same as termination. On a termination, a contractor may be entitled to expectation damages, or an owner may be entitled to compensation for having the work completed by another contractor. Rescinding a contract entitles neither party to anything. To rescind a contract is to make it void, as if it never happened. However, once a contract is rescinded, a contractor can seek restitution for any improvements made to the owner's property. Restitution for unjust enrichments will not ordinarily include the cost of estimating and bidding for the contract though because estimates and bids do not become valuable property that can be disgorged from the owner through restitution.

Mistakes Noticed After Award

A contractor who alleges a bid mistake after award must produce clear and compelling evidence to establish that a mistake was made. Suitable evidence includes the contractor's file copy of the bid, the original worksheets and other data used in preparing the bid, subcontractors' and suppliers' quotations, if any, published price lists, and any other evidence that establishes the existence of the mistake, the manner in which it occurred, and the bid amount actually intended.

If the evidence establishes that a bid mistake was made and correcting the mistake would be "favorable to the government without changing the essential requirements of the specifications" the contract may be modified to correct the mistake (FAR, § 14.407-4(a), 1984). But if evidence establishes that a bid mistake was made and correcting the mistake would <u>not</u> be favorable to the government, a mistake must be deemed to be either: 1) a mutual mistake; or 2) it must be a bid mistake that was, "...so apparent as to have charged the contracting officer with the notice of the probability of the mistake" (FAR, § 14.407-4(c), 1984). If a mistake is deemed neither a mutual mistake nor an apparent bid mistake, no change will be made to the contract. But if a mistake is deemed either a mutual mistake or an apparent bid mistake the government has the options to rescind the contract or to reform the contract. Reformation, however, will only occur, "...(i) to delete the items involved in the mistake or (ii) to increase the price if the contract price, as corrected, does not exceed that of the next lowest acceptable bid under the original invitation for bids.

Bid Mistakes by Subcontractors

Whenever federal procurement law governs a subcontract, the rules within the FAR will be applied to a subcontractor's bid mistake (*Sulzer Bingham v. Lockheed*, 1991). However, the ability of a subcontractor to withdraw its bid may be constrained by **promissory estoppel**. Promissory estoppel is a legal doctrine founded upon reliance. It can be applied to prevent a subcontractor from withdrawing its bid. It cannot, however, be applied to prevent a prime contractor from withdrawing its bid. The reason for these contrasting results is judicial reasoning to the effect that: (1) prime contractors can suffer financial harm for relying on the low bid of a subcontractor if that subcontractor is allowed to withdraw its bid; whereas (2) a public agency suffers nothing other than the loss of an unearned bargain if a prime contractor is allowed to withdraw its mistaken bid.

Bid Protests to Public Agencies

A **bid protest** is a challenge to a public agency's actions during the solicitation and award of a public construction contract. A person or business entity must have standing to make a bid protest. **Standing** is the legal right to make a legal claim or seek judicial enforcement of a duty or right (Garner, 2006). Standing is established in different ways, depending on the type of legal action and where it will be filed. For bid protest of federal procurement actions, standing is granted only to an "interested party." The FAR defines **interested party**, as follows:

> "Interested Party for the purpose of filing a protest means an actual or prospective offeror whose direct economic interest would be affected by the award of a contract or by the failure to award a contract." (FAR, § 33.101, 1984)

The FAR defines "bid protest," as follows:

> "Protest means a written objection by an interested party to any of the following:
>
> (1) A solicitation or other request by an agency for offers for a contract for the procurement of property or services.
> (2) The cancellation of the solicitation or other request.
> (3) An award or proposed award of the contract.
> (4) A termination or cancellation of an award of the contract, if the written objection contains an allegation that the termination or cancellation is based in whole or in part on improprieties concerning the award of the contract." (FAR, § 33.101, 1984)

Disappointed bidders are motivated to file bid protests in hope of disallowing a winning bid as nonresponsive, disqualifying it as not responsible, denying it for a bidding defect, or causing prices to be corrected with the end result of putting the protestor in the position of being the lowest responsible and responsive bidder. Public agencies are also vulnerable to bid protests for manipulating their choice of alternates in ways that favor a certain bidder.

A bidder may file a protest directly with the federal agency that is the object of the protest. Federal agencies have a **Board of Contract Appeals**, whose functions include, among others, adjudicating bid protests. If a bidder is reluctant to file a protest with the Board of Contract Appeals of the federal agency, the **Government Accountability Office [GAO]** provides an alternative. Federal bid protests may be filed with the GAO instead. The federal court system provides another alternative. Federal bid protests may also be filed with the **Court of Federal Claims**.

The protesting bidder chooses the venue: the Board of Contract Appeals of the federal agency, the GAO, or the Court of Federal Claims. The Board of Contract Appeals of each federal agency is likely in the best position to interpret construction contracts with that agency because its adjudicators will have specific construction experience. The GAO is one step removed from federal agencies but its adjudicators will lack the construction experience that is specific to each federal agency. The Court of Federal Claims is easily a neutral venue but where the agency and the GAO will adjudicate a claim in weeks or months, cases may take years to resolve within the Court of Claims. Whatever the choice, should the bidder be dissatisfied with the decision of the Board of Contract Appeals, the GAO or the Court of Federal Claims, an appeal may be filed within the Federal Circuit Court of Appeals.

The FAR encourages all parties to, "…use their best efforts to resolve concerns raised by an interested part at the contracting officer level through open and frank discussions…" before filing a bid protest (FAR, §33.103(b), 1984). But failing to resolve the issues, the protestor may file a written bid protest.

The protestor's filing must put forth the detailed, complete, and factual grounds for the bid protest. It must also include evidence to establish that the protestor is an interested party. Each agency will have its own filing requirements for inclusion of such things as the identities of the parties, the particular contract and clause at issue, and supporting documents.

When a bid protest alleges apparent improprieties in a solicitation, it must be, "…filed before bid opening or the closing date for receipt of proposals" (FAR, §33.103(e), 1984). In all other cases, protests must be filed no later than ten days after the basis of protest is known or should have been known, whichever is earlier (FAR §33.103(e), 1984). A federal agency will hear untimely filings only, "for good cause shown, or where it determines that a protest raises issues significant to the agency's acquisition system…" (FAR §33.103(e), 1984). Bid protests filed with the GAO must meet similar timeliness requirements. It behooves a protestor to be timely with its filings.

The Court of Federal Claims will hear protests filed within a six-year statute of limitations. However, this seemingly relaxed time limitation does not benefit a protestor. An ongoing construction contract will not be reassigned (i.e. a contract novation) to a successful bid protestor. And an agency, citing urgent public need, may even award a contract despite a pending bid protest. Once construction starts, the best a successful protestor can do is to recover cost of bid preparation and, if specific conditions are met, certain legal fees and costs. With great urgency then, bid protestors seek to prevent award, or to postpone notification to proceed, pending resolution of their bid protest. Because once construction starts, the practical relief available to a protestor becomes limited.

Chapter 12 SUBCONTRACTOR PROTECTIONS & VULNERABILITIES

12.1 Introduction

Subcontractors are essential to modern construction. They are indispensable for many reasons. They perform work that a prime contractor is unable to perform or work that a prime contractor chooses not to perform. A prime contractor may not be licensed to perform a particular type of work, such as electrical work or HVAC. Subcontractors are relied upon to perform such specialty work. But even when a prime contractor is licensed and chooses to self-perform a particular type of work they may still employ a subcontractor to take on work that is beyond their capacity to perform. When taking on work in a new, foreign, or unfamiliar market it is prudent for a prime contractor to employ local subcontractors. Local subcontractors are already familiar with that market.

Complex equipment may have its warranty tied to installation by a subcontractor who is approved by and licensed to install a particular manufacturer's equipment. Subcontractors are relied upon to install specialty equipment. Many subcontractors perform proprietary, licensed, and highly specialized work for prime contractors.

It is a rare project that does not employ subcontractors. Typically, subcontractors will perform 20% to 80% of the work. Some large prime contractors choose not to self-perform any of their work. Subcontractors working under their direction do it all.

Assignment vs. Subcontracting

To assign a task is to transfer it from one person to another. To subcontract a task is to entrust another with performing that task. These definitions for assigning and subcontracting, although very similar on the surface, have important legal distinctions underneath. Duties and rights under a construction contract cannot be assigned from prime contractor to subcontractor. A prime contractor can, however, subcontract performance to a subcontractor. Assigning and subcontracting are different when viewed through the eyes of the owner. A prime contractor cannot employ a subcontract to avoid liability to the owner for performance under his contract. If a subcontractor fails to perform, the prime contractor has failed to perform as far as the owner is concerned. The prime contractor's duty to perform cannot be assigned under the construction contract.

Having said that, there is actually a legal device for transferring the duties and rights of a prime contractor to a different contractor. This device is called a **novation**. A novation, however, requires agreement of both parties, both prime contractor and owner. The problem is that an owner is unlikely to agree to a novation. A prime contractor is on the hook to perform, no matter if nonperformance was the fault of a subcontractor, and an owner is not going to release that hook through novation.

Managing Risk

Good business people learn how to manage risk. The construction industry is filled with good business people. One of the principle ways that prime contractors manage performance risk – performance risk refers to the uncertainty associated with doing the actual work of the contract – is through subcontracting. An old construction maxim says, "Give the work to the subcontractor who is best able to manage the risk." So you would, for example, give electrical work to an electrical subcontractor who is licensed, more experienced than your own crews, or crews that you could train, and more likely to substantially complete the electrical work to the satisfaction of the owner. This is good performance risk management.

Another way that prime contractors benefit from subcontracting is that subcontractors typically will bill once a month. On receiving a subcontractor's monthly billing the prime contractor will generally have additional time to pay. It is not unusual for 6 to 8 weeks elapse between the time when the subcontractor's tradespeople first start work on a job and the prime contractor has to spend any cash to pay them. In contrast, when a prime contractor self-performs he has to pay his employees every week. The self-performing prime contractor spends cash much more quickly. The postponement of cash spending that occurs through subcontracting is referred to as **trade financing**, a term that alludes to the fact that the subcontractor is, in effect, "financing" the cost of labor for the construction project.

SUBCONTRACTOR PROTECTIONS & VULNERABILITIES

Communication: Contractor/Subcontractor Relationship

Most construction contracts will establish a single channel of communication between the prime contractor and the designer. Communication between any subcontractor and the designer is discouraged if not prohibited outright in the prime contract and the various subcontracts. By restricting communication in this way, the owner hopes to improve project outcomes by removing sources of miscommunication between the parties. As a practical matter, both the designer's representative and the owner's representatives will be talking to the subcontractors. Nevertheless, contracts will usually stipulate that all formal communications must be put into writing and that those writings must be exchanged between designer and prime contractor.

In a similar fashion, communication between the prime contractor and the owner is discouraged if not prohibited outright in the design contract. As a practical matter, the owner's representative will be talking to the prime contractor's representative. Nevertheless, contracts will usually stipulate that all formal communications must be put into writing and that those writings must be exchanged between owner and designer.

A subcontractor has two obstacles in the path of any communications with the owner. One obstacle is the prime contractor. The other obstacle is the designer. The subcontractor's communications must pass through both. So when a subcontractor has an issue, whether it be interpretation of the contract documents, standards of performance for the work, material substitutions, or a myriad of other issues that are encountered along the way, the prime contractor must step into the subcontractor's shoes and advocate for the subcontractor with the designer(s) and/or the owner. The subcontractor cannot communicate directly with either the designer or the owner. He must communicate through the prime contractor.

The prime contractor has various responsibilities toward his subcontractors. The prime contractor must provide competent coordination and supervision between the various subcontractors on the project, so that each subcontractor is enabled to perform their work. The prime contractor must timely pay approved invoices for work performed by its subcontractors. The prime contractor must establish a mechanism for communication and exchange of information between the various subcontractors, the designer(s), sub-consultants of the designer(s) and the owner. Finally the prime contractor must be his subcontractor's advocate for any disputes.

Each subcontractor has various responsibilities toward the prime contractor. The subcontractor must timely perform his work. All subcontractors must cooperate and coordinate with other subcontractors, the prime contractor, the designer(s), and sub-consultants of the designer(s), and the owner. A subcontractor must timely notify the prime contractor of potential claims. Finally, each subcontractor must adhere to contractual processes.

12.2 Subcontracting Issues

A subcontract defines the relationship between the prime contractor and the subcontractor in much the same way as the prime contract defines the relationship between the owner and the prime contractor. The substance of both types of contracts covers the responsibilities of the owner, contractor, subcontractors and designer; payments and completion; concealed or unknown conditions; changes; time; interpretation of contract documents; suspension and termination; uncovering and correction of work; claims and dispute resolution; insurance; and surety bonds.

Flow-thru Clauses and Conflicting Documents

In addition to the aforementioned matters of substance, it is common practice to include a **flow-thru clause**, sometimes referred to as a conduit clause, in the prime contract. A flow-thru clause compels the prime contractor to bind each subcontractor to the terms and conditions of the prime contract. Its purpose is to assure that the work of the contract is performed, without any gaps, exactly as specified within the contract documents. Typically, the flow-thru clause will extend both duties and rights to the subcontractor. It obligates the subcontractor with responsibilities to the prime contractor in the same way that it obligates the prime contractor with responsibilities to the owner. It then grants the benefits of rights, remedies and redress to the subcontractor in the subcontract in the same way that benefits of rights, remedies and redress are granted to the prime contractor in the prime contract. It will also protect the rights of the owner and designer with each subcontractor in the same way that the rights of the owner and designer are protected with the prime contractor.

A flow-thru clause is not self-executing. That is to say that merely appearing in the prime contract does not mean that subcontractors are automatically bound by it. The prime contractor must alert his subcontractors to the existence of a flow-thru clause. Typically, this is accomplished through **incorporation by reference** in each subcontract. Incorporation by reference makes the prime contract an integral part of the subcontract itself. A typical incorporation by reference subcontract clause states that the contractor and subcontractor are mutually bound by the terms of the prime contract and it extends duties and rights in the subcontract in the same way that they are assumed in the prime contract. Ordinarily there is a caveat expressing that in the event of a conflict between the prime contract and the subcontract, the subcontract terms control.

Typically, subcontractors are bound only to the extent of the work that they perform. There will also usually be an exception stating, "...unless specifically provided otherwise in the subcontract agreement," that enables a prime contractor to change the duties and rights of a subcontractor so that they are not exactly the same as those prescribed in the prime contract. Subcontractors can usually request copies of the prime contract to review before signing their subcontract and they might have the right to have terms and conditions in their subcontract identified wherever they are at variance with the prime contract. The flow-thru clause may also require each subcontractor to enter into similar agreements with each sub-subcontractor.

Prime contractors are prone to make certain subcontract provisions, such as payment provision, more stringent in their subcontracts than those that they enjoy in their prime contract with the owner. This is allowed under the "unless specifically provided otherwise" clause. Prime contractors do not normally make subcontract provisions less stringent.

Issues can arise when there is a genuine conflict between the terms and conditions of the prime contract and the terms and conditions of a subcontract. The "in the event of a conflict" clause is employed to resolve that issue. Typically, the conflict clause will have the subcontract govern over the prime contract.

Problems arise when the subcontract is silent on a provision that is addressed in the prime contract. The subcontract may be silent about dispute resolution, for example, while the prime contract includes an arbitration agreement. Does a flow-thru clause create an arbitration agreement between prime contractor and subcontractor? The Federal Arbitration Act requires agreements to arbitrate to be in writing between the parties. At least one court has ruled that incorporation by reference to a prime contract with an arbitration agreement creates a sufficient writing with which to enforce arbitration as between a prime contractor and subcontractor (Maxum v Salus, 1987).

Owner Selection and Approval

The owner may want to influence the prime contractor's selection of subcontractors. To that end, the prime contract may instruct the contractor to submit the names of the subcontractors, sub-subcontractors, and suppliers who will perform the principal portions of the work. Submittal is usually made to the designer in writing. The owner and designer may also assert a contractual right to object to any particular subcontractor. Generally, when the owner or designer objects to a particular subcontractor, the prime contractor is barred from subcontracting with them and the prime contractor must propose another. However, the prime contractor is usually excused from having to subcontract with anyone that it objects to.

The prime contractor will want a change order when it costs more to subcontract with another subcontractor. A deductive change order is implied if new costs are lower, but for obvious reasons, prime contractors will seldom submit a deductive change. Entitlement to an additive change order usually hinges upon whether or not the rejected subcontractor, sub-subcontractor or supplier was *"reasonably capable of performing the work."* To the extent that *"reasonably capable"* can be ascertained, a prime contractor is entitled to a change order only if the rejected subcontractor had been reasonably capable of performing the work despite the owner's or designer's objection.

Some owners will insist that the prime contractor self-performs part of the work. To that end the prime contract may limit the amount of subcontracting to a certain maximum percent of the contract dollar amount. Owners might also limit subcontracting to certain portions of the work. By that means a prime contractor who has expertise at a particular trade may not delegate that work to a subcontractor.

Public procurement officers at the state level and even some private businesses have started to use the construction procurement process as a means of achieving non-financial objectives. These objectives militate toward favoring subcontractors who reside in the location where the work is being performed; to suppliers and subcontractors from economically depressed areas; or toward environmental sustainability, workplace safety, civil rights, or protection of American manufacturers. These various objectives, meritorious as they may be, impose additional constraints upon intrinsically complex subcontracting processes.

Federal construction projects and public projects that are financed wholly or in part by federal dollars are subject to the Davis-Bacon Act, which requires that prevailing (union) wages are paid to all workers. Federally influenced projects may also be subject to affirmative action policies requiring certain minimum percentages of subcontract work to be awarded to small business enterprises and disadvantaged small business enterprises.

Bargaining Disparity

He who controls the scarce resource holds the bargaining power. This describes the ethos of subcontractors. Upon being awarded a prime contract with an owner, the prime contractor holds the bargaining power because he controls a scarce resource, a job. All of the subcontractors want a part of the job and a subcontract is the only way to get it. The more commoditized the subcontractor's service, the more this is true. Earthwork, concrete flatwork, framing, masonry, drywall are often treated as commodities to be awarded to the low bid subcontractor. This sort of commodity subcontracting is highly competitive. Prime contractors tend to award commodity subcontracts on a take-it-or leave-it basis where the subcontractor has no opportunity to negotiate more favorable terms.

A one-sided contract offered on a take-it-or-leave-it basis is known as a **contract of adhesion**. A subcontractor may feel like having been taken advantage of, but once signed, a contract of adhesion is a fully enforceable agreement, the law treating both parties as having negotiated a mutually agreeable contract even if negotiation was absent.

Few subcontractors are selling commodities any more. Today, they are selling specialized services. This arises because of the increasing complexity of materials, systems and methods. In 1950, if you needed glazing there were only two practical choices, 3/32" crystal or ¼" plate. Your glazer would be hired as a commodity subcontractor. Today, the choices are bewildering in their complexity. You would need a thick catalog to identify all of the performance characteristics of your potential choices for tinted glass, reflective glass, tempered glass, heat-soaked glass, heat-strengthened glass, insulated glass, low-e glass, laminated glass, fire-resistant, acrylic, polycarbonate and intumescent glass, to name just a few. Specialty subcontractors have proprietary knowledge, licensed methods, certified materials, warranty terms and the capability to provide post-construction services. The balance of power in subcontract negotiations has been tilting toward the specialty contractor. The more specialized its services the more this is true.

Bargaining disparity is highly evident in negotiations with suppliers of sophisticated equipment and specialized materials. Here, the subcontractor is pitted against large industrial manufacturers. A purchase orders for materials supplied by sophisticated manufacturers is likely to be a contract of adhesion. Unfortunately for the specialty subcontractor, it is he who must take-it-or-leave-it.

The relative bargaining power of any subcontractor could be placed on a continuum with the commodity subcontractor at one extreme and the specialty contractor at the other. Most subcontractors are going to fall somewhere in-between those extremes. These subcontractors in the middle, the majority of subcontractors, despite trending away from commoditization and into specialized services, still find themselves at a bargaining disadvantage against the large prime contractors. And even the highly specialized subcontractor is going to be at a bargaining disadvantage against the large industrial corporations, whom subcontractors rely upon for sophisticated equipment, systems and methods. All things considered, subcontractors suffer from a bargaining disadvantage.

Subcontractor associations have acted on the problem of bargaining disparity. The successful lobbying of various state legislatures by subcontractor's associations has resulted in statutory laws that mitigate some of the more serious consequences on both public and private projects. Statutes that protect subcontractors seek to maintain construction employment, a vital part of our economy.

Delayed Payment and Non-Payment

Most standard form construction contracts will call for both progress payments from the owner to the prime contractor and progress payments from the prime contractor to the subcontractor. Progress payments flow from owner to prime contractor to subcontractor. The separate contractual processes that make this happen are linked. When a prime contractor includes subcontractor's applications for payment in his own application for payment, and then receives payment from the owner, he is expected to disburse the appropriate sum to each subcontractor. Usually, the contract will stipulate that the subcontractor must be paid within a fixed number of days after the prime contractor receives payment from the owner.

If the prime contractor disapproves a subcontractor's application for payment the subcontractor will be notified and will not be paid until the deficiency is cured. Likewise, if the designer does not approve of the prime contractor's application for payment and cites the work of a particular subcontractor, that subcontractor will be notified by the prime contractor and will not be paid until the deficiency is cured.

Some states will intervene in both private and public construction contracts through regulation of the progress payment process. In California, for example, prime contractors must pay their subcontractors within 10-days of receiving payment from the owner and the subcontractor is entitled to a penalty of 2% per annum for any past due amounts.

Different things can happen when the owner, for reasons not the fault of the subcontractor, does not pay his prime contractor. Contract terms that are favored by subcontractors require the prime contractor to pay the subcontractor on demand, even when the owner has withheld payment from the prime contractor, as long as the subcontractor was not at fault for the withheld payment. Subcontractors do favor the usual contract terms: they require the subcontractor to wait for payment despite his lack of fault.

Pay-when-Paid Clauses

A contract clause that requires a subcontractor to wait for payment whenever the owner, for any reason not the fault of the subcontractor, does not pay his prime contractor is known as a **pay-when-paid** clause. Pay-when-paid is a harsh consequence for a subcontractor to bear but prime contractors have the bargaining power to impose the clause.

In most states a pay-when-paid clause will be enforced too, as long as it does not become a **pay-if-paid** clause. Pay-if-paid refers to a situation where the subcontractor never gets paid for work that he properly performed. Most states will expect a subcontractor to share the risk of delayed payment but not necessarily the risk of nonpayment. The subcontractor may have to wait for payment but the wait cannot be unreasonable. Most courts will enforce a pay-if-paid clause only as long as the subcontractor's wait for payment is not indefinite.

Retention

Retention, or retainage, is a set percentage of each progress payment that is withheld and accumulated by an owner as a contractual condition for assurance that the punch list will be completed and all required documents submitted upon final completion. A prime contractor will ordinarily assess retention from all subcontractor progress payments, at the same percentage that the owner is assessing the prime contractor. Some prime contracts will bar a prime contractor from retaining a larger percentage from a subcontractor than the owner retains from the prime contractor.

Some states will regulate retention processes. For example, in California a prime contractor must pay back all withheld retention to his subcontractors within ten days of receiving retention from the owner. If a subcontractor is not paid within the required time he is entitled to a penalty in the amount of 2% per month on the amount unpaid.

But even the California statutes will not help a subcontractor when there is a long, lawful delay in retention repayment to the prime contractor. Accumulated retention is typically paid back to the prime contractor upon final completion of a project. Many months or even years can elapse between the start of a project and its final completion. This long time lapse impacts any subcontractor who performed his work at the beginning of the project, such as an earthmoving subcontractor. To avoid this, subcontractors will try to get what is known as **line-item retention**. Line-item retention breaks up the work into separate line-items, each with a different retention schedule. With line-item retention, a subcontractor can get paid his retention without having to wait an unreasonable amount of time after satisfactorily completing his work.

Changes Clause in Subcontractor Agreements

Suppose that a prime contractor and an owner have executed a good, written agreement, and that agreement includes a changes clause. During the course of the work the owner orders a change that removes some of the work of an electrical subcontractor. The prime contractor can find himself in a difficult spot if the electrical subcontractor had started work without signing his subcontract agreement. Because while the prime contractor has no choice but to instruct the electrical subcontractor not to perform that part of the work – the prime contractor would be compelled to do that by the changes clause in his agreement - the electrical subcontractor cannot be compelled to reduce his price without his consent to the change. Without the electrical subcontractor's consent to a change, the court will enforce the electrical subcontractor's original basic bargain. The practical consequence of this is that the prime contractor will have to pay his electrical subcontractor the originally contracted for price regardless of how much work is removed from his scope.

Look for a changes clause in your written contract. If there isn't one and you are still negotiating the contract, try to get one inserted. Be sure to add instructions for how to handle deductive changes. Deductive changes are changes that remove items from the scope of work or otherwise reduce the cost of the work. Many a claim involving deductive changes hinges on indirect field costs, general overhead costs, and profit. Should you give some portion of each of those back to the other party on a deductive change? Or not? It is better to decide before work starts. Whatever it is that you and the other party agree to, insert it into your contract. Put changes clauses in your subcontracts. Make sure that all of your contracts are signed, particularly your subcontracts, before work begins

12.3 Mechanics' Liens

A **mechanics' lien** is a property right whereby the lien holder acquires the means to sell property and to be paid out of the proceeds. The sale is accomplished through a judicially supervised foreclosure. The property right is called a **security interest.** The opportunity to obtain a security interest through a mechanics' lien is created by statute. Every U.S. state (and Canadian provinces) has mechanics' lien statutes. Mechanics' lien statutes do not exist outside of the U.S. and Canada.

Most mechanics' lien statutes create a security interest only in real property not in other types of property such as personal property or intellectual property. Real property consists of land and things attached to land. Security interests cannot be filed against publicly owned real property such as state or federal buildings, highways and other infrastructure, or such things built by government agencies. Mechanics' liens are viable only with projects involving privately owned real property.

The purpose of mechanics' lien statutes is to encourage construction so as to maintain construction employment and motivate suppliers to provide materials for construction, stable employment and the sale of goods being vital components of every state's economy. Lien statutes engage that purpose by entitling prime contractors, subcontractors, suppliers, laborers and others to claim a security interest when in the course of construction they are not paid despite having earned the right to be paid.

Lien claimants can be thought of as arranged into tiers where the prime contractor and designer comprise the first tier; subcontractors, suppliers of the prime contractor, and sub-consultants of the designer comprise the 2nd tier; sub-subcontractors, suppliers of subcontractors, and sub-sub-consultants comprise the 3rd tier; and so forth.

Liens may arise from 1st tier claimants but 1st tier claimants have contracts directly with the owner. In the event of payment delay or failure they have the ability to look to their contract for relief. Parties in the 2nd and lower tiers do not have a contract with the owner. In the event of nonpayment their only contractual remedy is tied to their contracting partner against whom they can seek a court judgment for damages. A court judgment, however, is of no value against an insolvent debtor. It cannot create any money where none exists. A mechanics' lien, on the other hand, is a security interest against property that has real value. Mechanics' liens are very effective in the hands of a lower tier contractor.

A very common scenario occurs where the owner has paid a 1st tier party, such as the prime contractor, on an application for payment, but that party, in financial trouble or filing for bankruptcy, either chooses not to or cannot pay parties in lower tiers. Similarly, perhaps the 1st tier party has paid his 2nd tier parties but one of them does not or cannot pay parties in lower tiers. Mechanics' lien statutes enable lower tier claimants to acquire security interests directly against the owner's property. This is a very powerful device for lower tier contractors who would otherwise have nothing but a court judgment. Lien laws are criticized for being unfair to the owners, however, for owners may be compelled to pay twice, once to their 1st tier contractor and again to satisfy lien claims from unpaid parties in lower tiers.

In many states, the list of claimants extends to design professionals, design consultants, insurers, sureties, attorneys, employment agencies and others. Becoming a claimant hinges upon whether or not the claimant's work caused an *improvement* to real property. So an architect may not be eligible as a claimant if nothing yet has been constructed despite that he was not paid for his planning and design work. An excavation subcontractor who did no more than set out grade stakes would not yet have improved the property sufficiently to be eligible as a claimant. And a supplier of masonry may have to prove that her bricks were actually set into the building's curtain wall, not just strapped to pallets on site, before becoming a claimant.

Mechanics' liens are not the sole source of security interests in real property either. Property sellers, construction lenders, and mortgagers may also have security interests. Unlike with mechanics' liens, however, these security interests are created by contractual agreement between the parties. Those with security interests, regardless of how they were created, encumber property and an **encumbrance** stays with the property even if the property were sold to another. Encumbrances are said to "go with" the property.

Unencumbered equity is the difference between the market value of a property and the sum total of all encumbrances on that property. The sum total of all encumbrances may, and often will, exceed the market value realized in a judicial foreclosure. This leaves the court with the task of allocating the money. Most states will allocate monies *pro rata* by class. Classes are established chronologically, with "the first in time being the first in line." Construction lenders are usually the first class in line because their security interests are established before notice to proceed with construction. Construction lenders usually have priority. If there is any money left after satisfying the construction lenders claims it will be allocated among the next class in line, which could be mechanics' lien holders. Some states will allocate *pro rata* among all mechanics' lien holders. Others may give priority to laborers. In still other states, prime contractors' liens are subordinated to liens of laborers, subcontractors and suppliers. When all is said and done, the proceeds of a judicial foreclosure may be inadequate to satisfy all lien claimants. A mechanics' lien is only as good as the owner's unencumbered equity in the improved property.

Stop Notices

Some states enable liens to be placed on funds of the owner. This type of lien, commonly known as a **stop notice**, allows prime contractors, subcontractors, suppliers and others to tie up construction funds of the owner.

Stop notices are typically placed on the funds of the construction lender. As with a mechanics' lien, the stop notice is only as good as the size of the fund that it is attaching to. But a stop notice, even against a modest fund, provides great leverage to the claimant: a single subcontractor or supplier can literally stop progress on an entire construction project. More importantly, stop notices can even be placed against the funds of public owners. In states were they are enabled by statute, stop notices are valuable tools: they are more effective than mechanics' liens.

Statutory Lien Processes

Processes for filing liens vary in every state and state laws are constantly being changed. Courts will strictly enforce lien processes too. Slip up on any step in the process, even a minor clerical error, and you will lose your right to file a mechanics' lien. So beware. Learn the basic concepts. Become familiar with your local lien laws before starting any work. And consult your lawyer about the specifics as soon as any issue arises. Having said that, let's now look at some common lien processes.

Preliminary Notice Subcontractors or suppliers may be required to serve notice on an owner that they are performing work or supplying materials that are subject to mechanics' lien or stop notice. Preliminary notice is neither a threat nor a warning to an owner. It is beneficial information. Preliminary notices enable an owner to assemble a complete list of every potential lien claimant in every tier. That way there can be no surprises on his project. Notice must be served on an owner within a certain number of days after a subcontractor begins work or a supplier delivers materials or that subcontractor or suppliers loses his lien rights. Claims are only valid for work done or supplies delivered within that certain number of days before the notice was delivered and anytime thereafter.

Lien Claim A lien claim is a formal court filing that describes the work that was performed or the material provided, the name of the claimant contractor, subcontractor or supplier, the name of the property owner, the address of the job or a description of the place where the work was performed, and the amount owed. A lien claimant <u>must</u> file the lien claim within a specified number of days after the last day that claimant's work was performed or the material was accepted. Alternatively, tolling of the specified number of days can start on the date that work was accepted by the owner or upon achieving substantial completion. Contractors should be on guard for owners who file early notice of substantial completion, as this serves to reduce the timeframe within which lien claims can be filed.

Notice of Mechanics' Lien A lien claimant may be required to deliver a notice to the owner to which the lien claim is attached. Generally, the person serving the notice must sign and provide a proof of service affidavit.

Perfecting a Lien A lien claimant <u>must</u> file a lawsuit to establish a lien foreclosure action. This lawsuit must be filed within a specified number of days after filing the lien claim. Filing of the lawsuit perfects the lien. Failure to perfect the lien in a timely manner will invalidate the lien.

Release of Lien At any time after the lien claim is filed the owner can have the lien released by paying the amount claimed. In some states the owner can release the lien by posting a lien bond. And, of course, when the claimant's lien foreclosure action proves unsuccessful the lien will be released. It behooves owners to pursue release of liens because lien claims are usually recorded with the county recorder of deeds. Lien claims "cloud the title" of the owner's property making it difficult to either sell or secure a permanent mortgage for the property.

Just to get a sense of the specific time frames involved in the lien process: California requires preliminary notice to be filed within 20 days; lien claims to be filed within 90 days; and lien foreclosure actions to be filed within 90 days of the date that the lien claim was recorded.

Lien Waivers

A **lien waiver** is a written document, signed by a potential lien claimant, under which the lien claimant gives up the legal right to place a mechanics' lien on property to the extent of the amount asserted in the lien waiver. **Partial lien waivers** cover portions of the work, usually coordinated with progress payments, whereas **final lien waivers** cover the entirety of work on a project. It is common industry practice to exchange partial lien waivers in receipt of progress payments and final lien waivers upon final completion of the work and in receipt of the final payment including all accumulated retention.

A contract of adhesion may compel contractors to waive lien rights as a provision of their contract. Such contracts, known as **no-lien contracts**, have started to be regulated by state legislatures.

Special Payment Considerations for Subs

In the event that the owner or designer learns that a subcontractor or supplier intends to place a lien on the project, the designer might withhold certificates for payment. Rather than have payment withheld, prime contractors may choose to obtain a payment bond.

Another way that an owner can cause subcontractors and suppliers to be paid is to issue a joint check. A joint check is a check made payable to both the prime contractor and a subcontractor or supplier. Both payees must endorse a joint check. One payee cannot cash the check without the endorsement of the other. A joint check is a simple way to assure that money paid to prime contractors for work done by a subcontractor or for material supplied by a supplier actually gets passed on and paid to the subcontractor or supplier

Some contracts will bar a prime contractor from diverting money received for subcontractors' work. If the owner pays the prime contractor for a subcontractor's work but the prime contractor challenges the quality or completeness of that work, the prime contractor should either pay the subcontractor or return the payment to the owner.

Some contracts will authorize the designer to provide certain information, on request, directly to a subcontractor, such as: percentages of completion; payment amounts applied for by the prime contractor; and actions taken by the owner or designer with respect to the work of the subcontractor. An owner may be entitled to demand evidence of payment to a subcontractor from a prime contractor who is suspected of failing to pay after having received payment for the subcontractor's work. If evidence is not timely provided, the owner may be entitled to contact the subcontractor directly. These are all rare instances of contractually approved direct contact between designer and subcontractor.

Some contracts will protect the owner from lien claims from subcontractors and suppliers by creating a trust of the monies received by the prime contractor for the work done by those subcontractors and suppliers. A trust is a property interest held by one person at the request of another for the benefit of a third party. Here, the property interest is the legal right to the money received by the prime contractor. It is held by the prime contractor at the request of the owner for the sole benefit of the subcontractor or supplier. In the event that the contractor goes bankrupt, this trust gives the subcontractors and suppliers a preference over other creditors. This preference makes them more likely to be paid. When a subcontractor or supplier gets paid it eliminates the need to file a mechanic's lien against the owner.

Chapter 13　Ethical Considerations for Constructors

13.1　Introduction

Notwithstanding the honor and dignity that ethical business conduct begets, there are at least three practical reasons for ethical conduct in contractor-subcontractor business relationships:

(1) FAR § 52.203-13 requires federal contractors to: have a written code of business ethics and conduct; to make a copy of the code available to each employee engaged in the performance of the contract; to exercise due diligence to prevent and detect criminal conduct; and otherwise promote an organization culture that encourages ethical conduct and commitment to compliance with the law (FAR, 1984). These requirements may be incorporated by reference into subcontract agreements, thereby imposing its requirements upon both contractors and subcontractors.

(2) The Sarbanes-Oxley Act has a profound impact on corporate governance and financial practices of all business entities, large and small (Sarbanes-Oxley Act, 2002). Among its many provisions, businesses must disclose whether they have instituted a written code of ethics or provide an explanation if they have not.

(3) Subcontractors rely on a general contractor to provide competent coordination and project supervision; to timely pay approved invoices; to provide a channel of communication with the owner, designer, and other subcontractors; and to advocate for them in disputes. Contractors rely on their subcontractors to cooperate and coordinate with other subcontractors; to timely notify them of potential claims; and to adhere to contractual processes. These interdependencies create business relationships that are founded upon trust. Unethical conduct is toxic to a trusting business relationship.

13.2 A Code of Ethics

A code of ethics establishes the core principles and values of a business. Business managers are expected to communicate their code of ethics, compel its application throughout their organization, and institute disciplinary actions for any ethical violations. By crafting or adopting a written code of ethics, constructors bring themselves into compliance with both the FAR and the Sarbanes-Oxley Act.

Although no two constructor's code of ethics are exactly alike there are certain ethical topics that are common to all: bidding for work; subcontractor and supplier relationships; capabilities and competencies; conflict of interest; safety practices, employment practices, and legal compliance; professional status; and enforcement.

Bidding for Work

Besides bid shopping and bid peddling – invariably proscribed as unethical practices – a code of ethics will proscribe actions that violate any laws and regulations that govern the competitive process. Important proscriptions include falsities, kickbacks, and collusion.

Falsities

FAR §42.214-4 **False Statements in Bids**, requires the following clause to be inserted in all federal bid solicitations:

> "Bidders must provide full, accurate, and complete information as required by this solicitation and its attachments. The penalty for making false statements in bids is prescribed in 18 U.S.C. 1001." (FAR, 1984)

The **False Statements Act**, 18 USC §1001 asserts that a federal contract bidder who:

> "...knowingly and willfully falsifies, conceals or covers up by any trick, scheme, or device a material fact, or makes any false, fictitious or fraudulent statements or representations, or makes or uses any false writing or document knowing the same to contain any false, fictitious or fraudulent statement or entry, shall be fined under this title or imprisoned..." (False Statements Act, 2014).

Imprisonment of up to eight years is possible. However, the burden of proof under the False Statements Act is high. False statements are more often prosecuted under the civil **False Claims Act** (2014). The civil False Claims Act has a lower standard of proof than the criminal False Statements Act. It does not impose criminal penalties but each false claim is subject to a penalty of from $5,000 to $10,000 – multiple claims can result in very large penalties - and the government is also entitled to recover three times actual damages, otherwise known as "treble damages."

Kickbacks

The **Anti-Kickback Act of 1986** prohibits kickbacks in connection with federal contracting. Kickbacks are defined by the Act, as follows:

> "…the term "kickback" means any money, fee, commission, credit, gift, gratuity, thing of value, or compensation of any kind which is provided, directly or indirectly, to any prime contractor, prime contractor employee, subcontractor, or subcontractor employee for the purpose of improperly obtaining or rewarding favorable treatment in connection with a prime contract or in connection with a subcontract relating to a prime contract." (Anti-Kickback Act, §8701(2), 2011)

FAR §3.502-2 explains that, "…the Anti-Kickback Act of 1986…was passed to deter subcontractors from making payments and contractors from accepting payments for the purpose of improperly obtaining or rewarding favorable treatment…" (FAR, 1984).

Criminal violations may result in punishment of up to five years in prison and fines of up to $25,000. Civil actions may also be filed. The Act does not apply to either commercial bids or to federal solicitations when the bid amount is less than $100,000 (Anti-Kickback Act, §8703(d), 2011).

The **Copeland Anti-Kickback Act** addresses a different type of kickback, prohibiting the inducement of kickbacks from contractor's employees, asserting as follows:

> "Whoever, by force, intimidation, or threat of procuring dismissal from employment, or by any other manner whatsoever induces any person employed in the construction, prosecution, completion or repair of any public building, public work, or building or work financed in whole or in part by loans or grants from the United States, to give up any part of the compensation to which he is enti-

tled under his contract of employment, shall be fined under this title or imprisoned not more than five years, or both." (Copeland Anti-Kickback Act, 1948)

Collusion

Collusion is prohibited by the Sherman Antitrust Act of 1890 and prosecuted by the Antitrust Division of the United States Department of Justice (Sherman Antitrust Act, of 1890 [Sherman Act], 1890). Violation of the Sherman Act is a felony with a maximum fine of $100,000,000 for corporations, or $1,000,000 for individuals and a jail sentence of up to 10 years. Fines may be doubled for egregious violations and victims may be entitled to restitution. Civil remedies are also available whereby treble damages (three times actual damages) may be recoverable. Collusion among constructors usually involves price fixing, bid rigging or market allocation ploys.

Price fixing is any agreement between competitors to restrict price competition among them. Price fixing takes on many forms but generally will involve establishment of standard pricing, pricing formulas, fee schedules, discounts, or credit terms that are common to all. Colluders will also institute some form of self-policing mechanism to assure compliance with their price fixing agreement.

Bid rigging is any agreement between bidders to allocate jobs among them. It is the most common form of collusion affecting public agency solicitations. The ways to rig bids are as varied as the imaginations of constructors. In a **bid suppression** ploy, competing bidders agree to withdraw bids or to refrain from bidding in order to leave a pre-determined member of their conspiracy with the award. In a **complementary bidding** scheme, sometimes called a "courtesy" or "cover" bidding scheme, all but the pre-determined member of a conspiracy agrees to bid too high or to submit nonresponsive bids. Sometimes the reward for suppressing one's bid or putting in a cover bid will be a lucrative subcontract with the winning bidder, leaving the conspirators to split the profits. In a **bid rotation** ploy, members of a conspiracy take turns being the low bidder. The point of all bid rigging schemes is to covertly inflate prices while giving the appearance of fair competition.

Market allocation is a scheme whereby competitors allocate geographical regions, different types of customers, or different types of work among them. As with price fixing schemes, market allocation restricts competition and usually involves some form of self-policing mechanism to assure compliance with their market allocation agreement.

Ethical Considerations

ETHICAL CONSIDERATIONS

Other Code of Ethics Topics

Subcontractor and Supplier Relationships

Nothing is more destructive of project success than poor subcontractor and supplier relationships. Prime contractors should not lose site of the interdependencies that they share with their subcontractors and suppliers. Each has specific responsibilities toward the others. The prime contractor must provide competent coordination and supervision between subcontractors and suppliers, so that each is enabled to perform its work. Subcontractors and suppliers should be attentive to contractual processes; timely perform their work; and cooperate and coordinate with other subcontractors, other suppliers, and the prime contractor. A prime contractor must advocate for its subcontractors and suppliers with respect to claims while subcontractors and suppliers must timely notify the prime contractor of all potential claims.

Perhaps no one matter is more important that payment processes. Subcontractors and suppliers must submit timely and accurate payment requests and prime contractor must timely pay approved invoices for work performed by their subcontractors and suppliers. Special consideration must be given to statutory regulations of payment terms, retention schedules, and pay-when-paid clauses.

Capabilities and Competencies

The capabilities and competencies of a constructor are its value propositions, for the lack of which owners have no reason to buy. The lowest bid is still too much to pay for incapable or incompetent construction work. Constructors should neither seek nor accept work that they are not qualified to perform, for lack of technical competence and experience. But even where a constructor enjoys technical competence and experience, its capabilities are not unlimited. Diligent attention to staffing is necessary to avoid overcommitting management and supervisory resources. Customers must be treated with honesty and integrity, while prices must be established that are commensurate with the services that the constructor has capacity to provide. The minimum standard for delivery of materials and services must be set at acceptable standards that are established by the construction industry in the locality where work is being performed.

Conflict of interest

Conflict of interest appear when a firm's operating managers or those in positions of authority on specific projects have a financial interest, family relationship, friendship, or cultural, religious or political predisposition with respect to a particular client that may be perceived to be contrary to the best interests of the firm. In terms of business ethics, it is of little importance whether contrary acts are actually carried out. It is the perception of potential for committing contrary acts that defines conflict of interest.

FAR §3.1101 provides a practical definition of a personal conflict of interest, as follows:

> "Personal conflict of interest" means a situation in which [a contractor's] employee has a financial interest, personal activity, or relationship that could impair the employee's ability to act impartially and in the best interest of the Government when performing under the contract. (A *de minimus* interest that would not "impair the employee's ability to act impartially and in the best interest of the Government" is not covered under this definition.)
>
> (1) Among the sources of personal conflict of interest are:
>   - (i) Financial interests of the covered employee, of close family members, or of other members of the covered employee's household;
>   - (ii) Other employment or financial relationships (including seeking or negotiating for prospective employment or business); and
>   - (iii) Gifts, including travel.
>
> (2) For example, financial interests referred to in paragraph (1) of this definition may arise from
>   - (i) Compensation, including wages, salaries, commissions, professional fees, or fees for business referrals;
>   - (ii) Consulting relationships (including commercial and professional consulting and service arrangements, scientific and technical advisory board memberships, or serving as an expert witness in litigation);
>   - (iii) Services provided in exchange for honorariums or travel expense reimbursements;
>   - (iv) Research funding or other forms of research support;

(v)   Investment in the form of stock or bond ownership or partnership interest (excluding diversified mutual fund investments);
(vi) Real estate investments;
(vii) Patents, copyrights, and other intellectual property interests; or
(viii) Business ownership and investment interests." (FAR, §3.1101, 1984)

Safety, Employment and Compliance

Jobsite **safety practices** are largely centered on the OSH Act (Occupational Safety and Health Act, 2014). Yet it is well established in law that employers have a duty to maintain a safe workplace. This duty persists even in the absence of the OSH Act. In a construction setting, owner, architect and contractor can all be considered "safety employers," directly or indirectly, of workers on site, including subcontractors, sub-subcontractors, consultants, sub-consultants, inspectors, and any other person present in the workplace in connection with the work. Constructors share common interests in workplace safety. As caring people, none of them wants serious injuries or loss of life on their project. As business people, they want to avoid financial losses wherever they can and manage their risks wherever they cannot. Safety practices are an area where business, morality, law and ethics intersect.

Employment practices must be responsive to employment law. Employment laws confer rights to be protected from discrimination, to enjoy equal employment opportunity, to form unions and engage in collective bargaining with employers, to have terms and conditions of employment that meet at least minimum standards, to have basic liberties respected, and to receive compensation for certain types of harm done by employers.

While employment practices must be responsive to employment law, researchers advise that the social contract of employment may have greater influence on the working environment than any written employment contract. The **social contract** may be defined as, "the expectations and obligations that workers, employers, and their communities and societies have for work and employment relationships." (Kochan, 2014) Employees want to know that their company has a vision and a strategy; they want their own responsibilities within that vision and strategy to be clearly spelled out; they demand the tools and authority to achieve those responsibilities; they expect recognition and rewards for their performance; and they want to work for a supervisor who understands and motivates them (Davies, Kilmann, Orlander & Shanahan, 2009). A constructor may observe all labor and employment laws to the letter but still provoke discontent among its workforce for breach of the social employment contract.

Ethics codes will avow employers to **legal compliance**, using phrases such as, "Constructor will not knowingly violate any law or regulation." Such a phrase is more consequential than it might appear. The word "knowing" is a word of art in law. In certain circumstance it has a rather narrow meaning, that of "actual" knowing. Much like the answer to a test question, either you know it or you don't. More often, knowledge means not just what was actually known but what should have been known. This standard, known as *scienter*, puts a constructor in the position of knowing all laws and regulations impacting construction work because as an educated and competent professional contractor he or she should know those laws and regulations. Therefore, he or she is charged with knowing them regardless of whether or not he or she actually knows them.

Professional Stature

Ethics codes will include a pledge to promote and protect the **professional stature** of constructors and to foster standards of honor and dignity within the construction industry. But what exactly is the professional stature of constructors? Studies on professional stature in the eyes of the public, place the architect/engineer high up but the constructor way down. The **U.S. Justice Department** and other government agencies focus significant resources on ferreting out bid collusion. Others disparage contractors, depicting contractor licensing laws as, "state sanctioned consumer fraud" (Sweet, 1997, p. 11). Such things reflect badly on contractors, of course.

But things are different among business people and property developers, where the stature of constructors and architects/engineers are starkly reversed. Architects/engineers are dismissed as dreamers, uninformed on pricing and unable to run work (Sweet, 1997. p. 39). Constructors, in contrast, are seen as competent and informed managers (Sweet, 1997. p. 39). Owners seeking to engage the managerial skills, adeptness at pricing, and access to proprietary technologies that constructors provide are increasingly involving construction managers early in their design and planning processes.

Public perceptions have not yet caught up with business perceptions, but public perceptions can change. Ethical business conduct is essential if public perceptions are to change. Ethical business conduct cultivates an environment within which professional constructors can raise their stature in the eyes of business, government and the public.

Ethical Considerations

Enforcement

A code of ethics may be nothing but words if not for enforcement of its provisions. No code of ethics is complete without establishing internal procedures for its enforcement. Internal procedures for handling any ethical infractions must be clearly communicated to all employees. These procedures should be reviewed and renewed on an annual basis, with employees being routinely updated on all changes.

Employees must be encouraged to report ethics violations. Providing for anonymous reporting to an independent third party or a confidential hotline can mitigate employee insecurity. It is common for firms to assign enforcement responsibilities on a senior manager who reports directly to the firm's president or CEO. This ethics officer is charged with maintaining the code of ethics, initiating all ethics investigations and resolving all issues. Transparency and timely communication with all employees is essential for maintaining the integrity of a code of ethics.

FAR §52.203-17 requires federal contractors to inform their employees of their whistleblower rights and remedies (FAR, 1984). **Whistleblower** rights protect employees of prime contractors and subcontractors against reprisal for providing incriminating information to the government. These protections are prescribed in 41 USC §4712 and include: (1) protection for disclosing information; (2) identification of government entities to receive confidential disclosures; (3) processes for filing complaints when discrimination against whistleblowers is suspected; (4) remedies for discrimination against whistleblowers; and (5) requirements for employees to communicate these protections and remedies to their employees (Pilot program, 2014).

Incriminating evidence with respect to federal contracts includes: gross mismanagement; gross waste of funds; arbitrary and capricious exercise of authority, inconsistent with the mission of the executive agency or the successful performance of an agency contract; substantial and specific danger to public health or safety; or violation of a law, rule, or regulation related to a federal contract, bid solicitation or proposal.

13.3 Bid Shopping and Bid Peddling

A prime contractor, having received all the bids from its subcontractors and having identified the lowest responsive and responsible bidder, might then goad the other bidders into undercutting its low bidder's price. A subcontractor might do the same with its sub-subcontractors and sub-subcontractors to their sub-sub-contractors, and so forth. This is known as **bid shopping**. A subcontractor can also initiate by first approaching the prime contractor and offering to undercut the low bid of any competitors. Likewise, a sub-subcontractor can initiate with its subcontractor and so forth. This is known as **bid peddling**. Regardless of who does it, bid shopping and bid peddling is unethical conduct.

A subcontractor has to invest considerable time, money and effort to produce a winning estimate. The unethical, subcontractor need not spend a single minute estimating. As long as a willing prime contractor will reveal its low subcontractor bid, the bid peddler can place a bid slightly lower and steal for its own advantage the estimating work of its betrayed competitor. If another subcontractor steps in to undercut the bid still further, the competition can degenerate into a reverse auction. These auctions stop only when the losing seller is found: the one who becomes motivated to perform at the lowest level. The consequence of bid peddling is invariable low quality and excessive change orders, claims and disputes.

Even if no subcontractor peddles its bid, a subcontractor can still be worried about being bid shopped by the prime contractor. Prime contractors can bid shop both before award, to reduce the amount of their bid, or post award, to reduce their costs and increase their profit. Subcontractors who are nervous about **pre-award bid shopping** might attempt to deter bid shopping by waiting until the very last minute to submit their bid. Prime contractors' last minute scrambling to assemble their bid is the principal cause of bid mistakes. Subcontractors who are nervous about **post-award bid shopping** might simply refrain from bidding. And the prime contractor who develops a reputation for post-award bid shopping will ultimately find that subcontractors inflate their bids in anticipation of the levy to be extracted from it.

Bid shopping and bid peddling are unethical but not unlawful. Subcontractor associations have sponsored legislation in the various states to prohibit bid shopping and bid peddling or to mandate bid depositories. However, all such statutes have been struck down by the courts as restraints against competition in violation of the Sherman Act.

Although bid depositories are not viable in the U.S., other nations and international construction projects have invoked them to good effect. A **bid depository** is a physical or virtual facility that is operated by a third party, not a participant in the bidding process. Subcontractor bids are submitted to the bid depository. The bid depository maintains a permanent record and forwards bids on to the appropriate prime contractors. Because the bidding records in the depository cannot be manipulated, they create a transparency that discourages bid shopping. Bid depositories are particularly effective when combined with listing: a requirement for prime contractors to list their low bidding subcontractors when submitting their bids and to actually use those subcontractors when performing the work.

Irrevocability of Subcontractor Bids

In the common law of contracts a subcontractor's bid is an offer. By submitting an offer a subcontractor grants the power to accept its offer to the prime contractor who receives it. A valid acceptance forms a contract. The prime contractor need not accept immediately because an offer remains open for a reasonable time. But a reasonable time is not very long: for a written offer it is the end of the day upon which the offer is received. So a bid expires at the end of the day upon which it is received. Thereafter, it cannot be accepted to form a contract.

A prime contractor, of course, will never accept a bid on the same day that it is received. It takes some time to evaluate subcontractor bids, to determine if they are responsible and responsive, and to create a bid-tabulation. Time-consuming negotiations are also common, to iron out any misunderstandings and to reinforce all understandings regarding the scope, schedule, price, and quality of the work. And of course the prime contractor wants to first win the job before committing to any subcontracts. To preclude acceptance problems, prime contractors might require subcontractors to render their offers irrevocable for a specified period of time, typically 30 to 90 days.

In the common law, however, rendering an offer irrevocable is in itself a contract because it confers a benefit that would not otherwise be present. As with all other contracts, consideration must be present for the contract to be enforceable by the courts. An irrevocable bid that is not supported by consideration is seen as a donative promise, unenforceable by the courts. In its effect, this means that unless a prime contractor pays a subcontractor to make its bid irrevocable - something that they will rarely pay for – a subcontractor can withdraw its bid at any time prior to acceptance, even if its bid states that it is irrevocable. This was the common law reasoning that was held by most courts in the past and a few courts yet today (*Baird v. Gimbel*, 1933).

Today, most courts think otherwise. Courts in a majority of the States will apply the doctrine of **promissory estoppel** as a substitute for consideration, thereby extending contract law to include irrevocable bids (*Drennan v. Star Paving*, 1958). In its effect, promissory estoppel bars a subcontractor from revoking its bid for a specified period of time.

The doctrine of promissory estoppel is founded upon reliance, reasoning that one who justifiably relies on another's promise should not suffer the consequences of the failure to make good on that promise. For promissory estoppel to apply in a subcontractor bid context, four conditions must be met:

1. The subcontractor's bid must form a clear and definite offer;
2. The subcontractor must reasonably expect that its offer will induce the prime contractor to rely upon it;
3. The prime contractor must actually rely upon the subcontractor's offer; and
4. Injustice can only be avoided by enforcing the offer.

All four conditions are met when: (1) a subcontractor submits the lowest responsive and responsible bid; (2) in response to the prime contractor's invitation to bid; (3) the prime contractor uses that low bid when forming its own bid, whereupon the prime contractor then wins the award at its bid price; and (4) the prime contractor would suffer a financial loss were the subcontractor to now withdraw its bid. Under these circumstances the subcontractor would be liable to the prime contractor for damages incurred by the prime contractor in reliance on the subcontractor's bid. Ordinarily that would be the difference in price between the subcontractor's bid and the next lowest responsible and responsive subcontractor's bid. Courts are not likely to compel a subcontractor to execute a contract with the prime contractor.

Courts in some states have mixed views on promissory estoppel, rejecting the doctrine in some cases while supporting it in others. A few states do not recognize it whatsoever. In those states, subcontractor bids are revocable at any time prior to acceptance.

Bid Shopping an Irrevocable Subcontractor Bid

When a subcontractor is barred from revoking its bid it sets up the prime contractor with a one-sided advantage. Confident in knowing that the subcontractor's low bid is locked in and irrevocable, the prime contractor is then free to aggressively bid shop competing subcontractors' bids.

Bid shopping might be avoided through presumptive acceptance. The term **presumptive acceptance** refers to language in a subcontractor's bid that states that acceptance occurs upon using the subcontractor's bid. Unfortunately, presumptive acceptance rarely succeeds in court.

Many states have enacted listing statutes. Directed at public construction contract solicitations, **listing statutes** require a prime contractor to list by name all of the key subcontractors that it will use in its work. Contractors are barred from using non-listed subcontractors, except for certain exceptional circumstances. Listing had at one time been embraced by the FAR but the practice has since been abandoned.

Subcontractors can avoid irrevocability by fashioning their price as a "quotation" or "estimate" provided for the prime contractor's convenience and not as a bid. However, such quotations might be determined to be non-responsive.

A prime contractor can always enter into contingent contracts with its low bidding subcontractors before receiving the award from the owner. If the prime contractor doesn't win the award its contingent contracts are extinguished, without penalty. But if the prime contractor wins the award, its contingent contracts become fully enforceable. Not only are contingent contracts highly ethical but they are also a great way for a prime contractor to win respect from its subcontractors

Reverse Bid Auctions

Some public agencies are soliciting for competitive bids through reverse bid auctions. A reverse bid auction is a controversial form of electronic bidding, or **e-bidding**, for public construction contracts. A reverse bid auction is the opposite of a normal bid auction. In a normal bid auction, buyers bid up the price on something until the bidder with the highest price wins. But in a **reverse bid auction,** sellers bid down the price that they are willing to be paid until the seller who accepts the least wins.

These auctions are conducted online using digital technologies. The auction is announced with an ITB. There may be a pre-bid conference conducted at the owner's facility or online. Bidders must submit responsive and responsible proposals. In all other respects a reverse bid auction is identical to other solicitations but for requiring bidders to bid prices online, rather than conveying a single price in a sealed envelope.

All bids are revealed online as they are made, with the intent of provoking bidders to undercut their competitor's bids. Undercutting of bids after the fact resembles bid shopping or bid peddling..

The American Subcontractors Association [ASA] has appealed to the U.S. Congress to: "Deter bid shopping and bid peddling by prohibiting the use of reverse [bid] auctions for construction and construction-related services at both the prime and subcontract levels." (American Subcontractors Association, Inc., 2013). The ASA, in its depiction of reverse bid auctions, is trying to convince Congress to stop public agencies from behaving unethically. Public agencies think otherwise: they contend that reverse bid auctions are in the best interest of public agencies (and by extension, the best interests of the public) because they promote competition. They argue that competition yields the lowest and best prices. The industry is clearly at odds with public agencies over reverse bid auctions and this controversy will no doubt continue.

Chapter 14   CONSTRUCTION INSURANCE

14.1   Introduction

Constructors risk monetary losses due to their performance on various construction contracts. Constructors also risk monetary losses from accidents that happen to people and property that have nothing to do with their construction contracts. Tower cranes have been known to collapse, causing damage to adjacent property and the injury and death of innocent bystanders. Heavy civil constructors that use explosives could inadvertently cause injury, death and damage to property far removed from their construction site. Highway accidents can happen on public roads caused by vehicles owned by the constructor and being used in the course of its business.

Whether or not a contractor is legally obligated to pay for accidents usually depends on whether or not the conduct of the contractor's employees met a standard of care to protect others from harm. Negligence is the failure to meet this standard of care. Negligence is one of a class of civil wrongs called torts. Put simply, a tort is a wrongful act. There are many different types of wrongful acts. The most commonly encountered wrongful act in design and construction, by far, is negligence. This chapter will have a special focus on negligence.

Negligent conduct is insurable as long as that conduct is not willful. Commercial general liability insurance relieves a contractor from monetary damages arising from its negligent conduct. Monetary damage to a contractor's work arising from explosions, fire, ice, snow, theft, vandalism and other casualties are insured with a special types of property insurance called builders risk insurance. Contractors also need equipment floater insurance, automobile insurance, umbrella excess liability insurance, and workers' compensation insurance.

A contractor cannot buy insurance to cover its performance on the work of its contract. Performance and re-performance of the work of its contract is the business risk that is assumed by the contractor. This risk allocation rule, known as the **work product exclusion**, is standard practice in the insurance industry. A contractor cannot buy insurance to cover defective materials or warranty work either. Faulty performance, defective materials and warranty work have to be corrected at the contractor's own expense.

14.2   Axioms of Liability

Negligence, put simply, is the lack of due care (Negligence, 1971, §1). Due care must be exercised to protect others from the unreasonable risk of harm. **Liability** is defined as a legal accountability to another party. When a plaintiff[1] files a lawsuit and the defendant[2] is found liable for negligence, the court will usually order the defendant to pay damages to the plaintiff in an amount that will compensate the plaintiff for the losses that he or she incurred as a direct consequence of the defendant's negligence.

To the layman, that a contractor is found liable for negligence would infer that the negligent contractor's employee(s) must have been partly, if not wholly at fault for whatever harm was done. But the legal standard for negligence is not based upon fault, at least not in the moral sense with which we normally comprehend fault. Morally, a person is not at fault if they did the best that they knew how to do, despite a bad outcome. The negligence standard, however, holds people accountable not for what they actually knew but for what they should have known.

Vicarious Liability

An agent is somebody who acts on behalf of another, known as the principal. Once agency is established, the principal becomes liable for acts of his agent that are committed in the course and scope of the agent's work. This type of liability is known as **vicarious liability**.

Employees such as tradespeople are agents of the contractor that employs them. Liability for the acts of employees will attach, vicariously, to the contractor who employs them. When we speak of a contractor's liability

---

[1] A plaintiff is the party who files a legal complaint. In construction law, the typical plaintiff will file a negligence suit against a contractor, owner, or another party to a construction project. Most plaintiffs are 3rd-party plaintiffs, meaning a party who is not in privity to any contract within the construction project from which the lawsuit arose.

[2] A defendant is the party who must respond to a legal complaint. In construction law, the typical defendant is a contractor, owner, or any other party to a construction project who is named in a negligence suit.

for negligence what we are really talking about is a contractor's vicarious liability for the negligent acts of its employees.

A designer does not want a contractor to become his agent because that would make the designer liable for the contractor's negligent acts. This is especially problematic with respect to jobsite safety. The designer could create an unintended agency by supervising the contractor's work or by prescribing construction means and methods. Means and methods have special significance with respect to the law of agency.

The contract documents will typically proscribe the designer from supervising and directing the work or having control over construction means, methods, techniques, sequences and procedures. This insulates the designer from vicarious liability for the contractor's acts and assures, among other things that the responsibility for jobsite safety remains with the contractor. It does not, however, preclude either the designer or owner from voicing safety concerns to the contractor.

Negligence

Every person owes due care to all other people. Due care is a standard of care that is common to all. The standard of care for a person who is a contractor is the competence that should be expected of a contractor. It doesn't matter if a contractor does the best that he or she knows how to do. If their work does not meet a standard of competence common to all other contractors, then their work can be found negligent. The legal standard of care for an architect or engineer is what a professional architect or engineer, working in the same geographical region, would do to protect others against unreasonable risk of harm.

Four elements must be proven to find negligence:

- Duty: The defendant had a duty to protect the plaintiff from unreasonable risk of harm;
- Breach: The defendant did not fulfill that duty;
- Cause: The defendant's conduct caused the harm that was suffered by the plaintiff; and
- Harm: A protected interest of the plaintiff was harmed.

CONSTRUCTION RISKS

Duty

It is usually enough to show that a defendant's conduct created an unreasonable risk of harm to establish that that defendant owed a duty to protect the plaintiff from that risk. However, the courts have the authority to reverse that determination and conclude that a defendant had no duty, when doing so is justified by an important legal principle or public policy. A contractor who uncovers unforeseen hazardous materials, for example, is clearly exposing many potential plaintiffs to an unreasonable risk of harm. That would be enough to establish a duty to protect those plaintiffs but for federal legislation that places liability for hazardous materials solely upon the owner and not the contractor.

Breach

That a defendant did not fulfill a duty, sometimes referred to as a breach of duty, can be established directly or indirectly. Establishing a breach directly can be done when the defendant is observed doing something harmful. Establishing a breach indirectly is established, as a general rule, when: the harm could not occur except for somebody's negligence; the defendant was in control of the instruments that caused the harm; and the plaintiff was not to blame. So, for example, a hammer falls from a roof where a contractor is doing work and that falling hammer strikes a passerby on her head, causing an injury. Even if nobody actually saw a roofer drop the hammer, negligence can be established with the following facts: A hammer would not fall from the roof by itself but for the negligent conduct of some roofer; all of the roofers on site were employees of the accused contractor; and the passerby had no way to know that roofers were working on the roof.

Cause

Establishing that the defendant's conduct caused the harm that was suffered by the plaintiff has two components: factual cause and proximate cause.

Factual cause is spoken of as the "but for" rule. In our example above a passerby was injured when struck in the head by a falling hammer. Let's now complete the story. An ambulance is summoned to pick up our passerby and deliver her to a nearby hospital, where a doctor, in the course of treating her injury, orders intravenous medication for an apparent concussion. She reacts adversely to the medication, however, and dies in the emergency room. Did our roofer's conduct cause her death?

Our roofer is the factual cause of our passerby's death. How does factual cause analysis determine this? Let's apply the "but for" rule. But for the falling hammer, she would not be injured. But for that injury she would not be transported to a hospital. But for being treated in a hospital she would not be given medication. But for being given medication she would not suffer an adverse reaction and die. By the "but-for" rule, our roofer's conduct is the factual cause of our passerby's death.

Yet it still must be shown that our roofer's conduct was the proximate cause of our passerby's death. **Proximate cause** sets limits on the types of consequences that liability may attach to. If not for proximate cause, liability could attach to every consequence, no matter how distant and remote it was. The key to proximate cause analysis is foreseeability. Liability extends to foreseeable consequences but not to unforeseeable consequences.

In our example, however, an adverse reaction to medication and death is a foreseeable, although uncommon, consequence of medical treatment. People die often enough from competent medical treatment that death is considered a foreseeable consequence. The crucial distinction here is that medical treatment is both necessary and risky. Proximate cause analysis concludes that our roofer's conduct proximately caused the death of our passerby. Under our concussion scenario the roofing contractor's conduct is both the factual cause and proximate cause of our passerby's death.

Let's change our example now. Let's say that on the way to the hospital the ambulance passes over a bridge, which collapses, dumping the ambulance into a river and drowning our passerby. The bridge was later determined to have had hidden structural damage. Is our roofer's conduct still the proximate cause of death? Proximate cause analysis would say that is was not foreseeable that the bridge would collapse at the very moment that our ambulance was passing over it. The bridge's collapse would be seen as proximately caused by hidden structural damage; its collapse an intervening event. Under our bridge collapse scenario the roofing contractor is the factual cause of death but not the proximate cause of our passerby's death.

Harm

Even when duty, breach and cause, both factual and proximate, are proven, for negligence to be found, a **protected interest** must be harmed. There are four types of harm that can qualify as protected interests: personal injury; property damage; **economic losses**; and psychological or emotional injury, sometimes called noneconomic losses. Every court will find that personal injury and property damage are protected interests. Accordingly, damages will always be awarded for the consequences of death, personal injury and damage to property on a finding of negligence by the courts.

Losses related to business operations (i.e. cost, profit, overhead and consequential damages), are known as economic losses. Most state courts will not recognize economic losses as protected interests in negligence suits (Sweet & Schneier, 2013, §4.10). A few courts have provided protection from economic losses in negligence suits but such outcomes are infrequent and unusual. Courts are even less apt to protect noneconomic losses. As a practical matter, when either economic or noneconomic losses are the consequence of construction activities an aggrieved party must look to its contract, if they have one, for a remedy because they will not find a remedy in negligence.

Strict Liability

Construction activities such as working with explosives, pile driving and demolition are often deemed **abnormally dangerous activities**. Contractors are held strictly liable for abnormally dangerous construction activities. **Strict liability**, put simply, means liability without fault. Strict liability for abnormally dangerous construction activities asserts that if someone is harmed, as a direct consequence of an abnormally dangerous construction activity, the contractor is liable. It doesn't matter if the contractor's employees conducted the activity with utmost care and took every possible precaution. The contractor is liable because the contractor voluntarily chose to perform an activity that is abnormally dangerous.

Statutory laws in most states prohibit strict liability from being delegated. Where the designer has specified work for which abnormally dangerous activities are unavoidable, liability cannot be delegated to the contractor who performs those activities. It would be irrelevant that the contractor or a subcontractor supervised and conducted the activities. Typically the owner, assumed to have ordered and approved the design, would be held strictly liable. The owner would be held strictly liable for having necessitated abnormally dangerous construction activities.

On the other hand, were the contractor to choose an abnormally dangerous activity, not because the contract documents dictated it, but for the contractor's purposes insofar as cost or construction efficiency, the contractor would be held strictly liable regardless of whether the work was delegated to a subcontractor or not.

Owners and contractors can insure against these liabilities. But it is important for both to recognize any abnormally dangerous activities that are necessitated by the contract documents or the contractor's work plans. By recognizing abnormally dangerous activities in advance they afford themselves the opportunity to either find suitable insurance or to change their plans to eliminate the need for abnormally dangerous activities. Both parties will want to include the cost of insurance, known as a risk premium, in their calculations of the price for the work.

Relief from Liability

Plaintiffs may seek relief in court even when the harm that they suffered was partly caused by their own conduct. The courts in the various states differ in the rules that they apply for awarding damages whenever a plaintiff is partly to blame for his or her own injury. And a defendant might be relieved from liability entirely if the plaintiff can be shown to have assumed the risk of injury. Finally, when a public owner such as a municipal, state, or federal government is to blame a plaintiff may be unable to file suit against that public owner.

Contributory Negligence

Alabama, Maryland, North Carolina, Virginia and the District of Columbia follow the **contributory negligence** rule. The contributory negligence rule asserts that a plaintiff who is partly to blame for his or her own injury cannot recover damages. Put simply, no matter how negligent a defendant was, if the plaintiff was even slightly to blame for his or her own injury then he or she is barred from recovering any damages whatsoever. This is a very harsh rule and juries in contributory negligence jurisdictions have been known to soften the harshest consequence of this rule. It is instructive to know that all jurisdictions in the U.S. followed the contributory negligence rule at one time or another but today all but these five jurisdictions have adopted more liberal rules.

Comparative Negligence

Alaska, Arizona, California, Florida, Kentucky, Louisiana, Mississippi, Missouri, New Mexico, New York, Rhode Island, and Washington follow the **comparative fault** rule. The comparative fault rule asserts that a plaintiff who is partly to blame for his or her own injury can recover damages, but the amount of that recovery is reduced by their share of fault. So when the court apportions fault 80% to the defendant and 20% to the plaintiff, the plaintiff's reward is reduced by 20%.

The remaining 34 states have adopted a **modified comparative fault** rule. The modified comparative fault rule asserts that a plaintiff who is partly to blame for his or her own injury can recover damages in an amount reduced by their share of fault, but only if the plaintiff's share of fault does not exceed a specified number. Arkansas, Colorado, Georgia, Idaho, Kansas, Maine, Nebraska, North Dakota, Tennessee, Utah, and West Virginia use a 50% bar rule. In these eleven states a plaintiff cannot recover if their share of fault is 50% or more. Connecticut, Delaware, Hawaii, Illinois, Indiana, Iowa, Massachusetts, Michigan, Minnesota, Montana, Nevada, New Hampshire, New Jersey, Ohio, Oklahoma, Oregon, Pennsylvania, South Carolina, Texas, Vermont, Wisconsin, and Wyoming use a 51% bar rule. In South Dakota a plaintiff cannot recover if their share of fault is more than "slight". Although South Dakota's rule is called modified comparative fault its effect is closer to that of the contributory negligence rule.

Assumption of Risk

Let's say that a tradesperson were to work for a contractor knowing that the work was very dangerous, that the contractor was not going to protect him or her from the risk of injury, but that tradesperson was willing to do the work despite the risk and regardless of whether the contractor conduct were negligent. Such circumstances define **assumption of risk**. The author has witnessed these circumstances in third-world countries were workers may have no choice but to assume all risks. If they are injured they are simply tossed off the job: the contractor is not liable; emergency medical treatment is not available; there is no medical insurance; injured workers are simply discarded like broken tools.

Most U.S. employers are barred from the assumption of risk in their employment relationships through express words in employment contracts or labor agreements if not by statute in the various states. Assumption of risk claims do sometimes arise between tradespeople and employers other than the one whom they are working for. Most of these claims assert that the assumption of risk was implied and courts will require the defendant to prove that the plaintiff impliedly assumed the risk. Claims of implied assumption of risk are rarely successful. But if assumption of risk can be proven a plaintiff cannot recover damages, even when the defendant was found to be negligent.

Sovereign Immunity

Sovereign immunity is a common law rule that shields a governmental body from being sued without first giving its consent. Congressional legislation long ago waived the federal government's sovereign immunity for negligence claims. There are exceptions in the federal statutes, however, that still protect the federal government and its agencies from being held liable under certain circumstances. Many of the states have also enacted statutes that enable state and municipal government bodies and their agencies to be sued for specific types of claims.

Construction companies who bid on public projects need to retain competent legal counsel in order to assess their risks with respect to their ability and that of third parties to sustain a lawsuit against any government body. Even where sovereign immunity has been waived, there will still be difficult obstacles to overcome: from strict time limitations for filing suits to requirements for legislative bodies to review each claim and approve any lawsuit before it can proceed.

Indemnification

Indemnification (alternately indemnity or to **indemnify**) is a legal device for shifting losses from one party to another. In the usual context of construction, "A" (the **indemnitor**) indemnifies "B" (the **indemnitee**) when "A" agrees to pay for any losses that "B" sustains from lawsuits against "B" or legal liability arising from the negligent acts or other torts committed by "B". "A" must also defend "B" in court and pay attorney's fees and court costs.

If a person or their property is harmed through negligence, that person or that property owner is going to want to be compensated for their losses. They could file a lawsuit. They might sue the contractor on evidence that the contractor's employees violated jobsite safety rules. Or they might choose to sue the designer realizing that the designer should have detected the jobsite safety rules violation but didn't. Then again they might choose to sue the owner knowing that the owner did not terminate the contractor after learning of the jobsite safety rule violations. Recognizing that the owner might have deeper pockets (e.g. more money) than either a contractor (contractors often have cash flow problems) or a designer (designers may have few assets) their lawsuit may be directed at the owner.

A person who is harmed can sue whomever harmed them and in whatever manner they choose. They can choose to sue one particular party, different parties in separate lawsuits, or multiple parties in a single lawsuit.

Construction Risks

If they win they can only collect one award, however. They cannot double-up or triple-up on their actual damages by suing multiple parties for the same claim. Aggrieved parties structure lawsuits to maximize the likelihood that they will be compensated for their losses. How they do that often has little to do with who on the construction project was actually to blame. It will have to do with who on the construction project has money. The losing party can end up paying the entire award despite having little or no blame.

Statutory laws of the various states will sometimes allow the sued party to determine who was actually to blame through special court proceedings that result in contributions from other parties. Other legal proceedings may result in distributions among culpable parties that are based not on statutory laws but on common law principles of equity. Some states allocate a *pro rata* share (e.g. three parties to blame so each pays $1/3^{rd}$). In other states each party contributes based on their comparative culpability (e.g. party A 50%, party B 30%, and party C 20%). The process of allocating losses to the parties actually culpable can be complicated, time consuming and very messy.

To avoid these allocation problems, the standard form construction contracts adopt **indemnification** language. Typically, the contractor is required to indemnify the owner and the designer, plus consultants, agents and employees of any of them, from losses caused by the negligent acts of the contractor plus their subcontractors, sub-subcontractors, suppliers and their agents and employees. In this context indemnify basically means to pay for damages awarded in lawsuits claiming negligent acts. The contractor must also defend the owner and/or the designer in court and that requires paying attorney's fees and court costs.

Indemnification language may not require the contractor to pay for all damages. It may only require the contractor to pay for damages in which the contractor was culpable and then only to a comparative degree of culpability. So if the contractor was 20% culpable it pays 20% of the damages. If 50% culpable it pays 50% and so forth. This comparative type of liability also applies to attorney's fees and court costs. This scheme is known as **narrow form indemnification** or **comparative form indemnification**.

Some indemnification agreements will require the contractor to pay 100% of all damages in which the contractor was at least partially culpable. This scheme is known as **intermediate form indemnification**. This contributory type of liability also applies to attorney's fees and court costs. Most states will enforce intermediate form indemnification language.

Indemnification agreements will not normally require a contractor to indemnify others for losses where the contractor had absolutely no blame in the loss. This indemnification scheme is known as **broad form indemnifi-**

cation. Statutes in some states prohibit broad-form indemnification. In those states, losses based on negligence solely by the owner or the designer must be defended and paid for solely by the owner or the designer.

Damage to the work itself is not covered by any of these indemnification schemes. Nor is damage due to failure of the contractor to construct the building in conformance with the contract documents. Legal counsel should review the indemnification language in the contract documents because indemnification statutes in the state where the work is being performed may change its effect. Indemnification language will ordinarily be altered so that it is not in conflict with statutes.

Subrogation

Subrogation is a legal principle that transfers a legal right from one person or entity to another. Subrogation legally substitutes an insurer in place of its insured policyholder. In common law the only person or entity with standing to file a lawsuit - if not for subrogation – is the person who was harmed. Only persons with standing have the legal right to sue. If the person who was harmed was insured, his insurance company pays damages pursuant to his or her claim. Insurers subrogate after paying a claim in order to enable them to file a lawsuit to recoup their money from whomever was legally responsible for the damages that they were obligated to pay.

Insurers cannot subrogate against the person that they insure. This rule bars an insurance company from paying a claim only to turn around and sue their policyholder in order to get their money back. If not for subrogation, the purpose of insurance would be defeated.

It is now common practice for contractors to identify the owner, designer, consultants, subcontractors, and sub-subcontractors as **named insureds**. By doing so they bar their insurance company from using subrogation to sue other project participants. Whenever the contractor is partly culpable, any named insureds that are also partly culpable need only be joined in the same lawsuit. Consolidating all defense costs under one insurance policy in this manner saves legal expenses for all insured parties.

The Fiduciary

A **fiduciary** is someone with a legal duty to act in the best interests of another or suffer the consequences for not doing so. Common examples of fiduciaries include trustees, legal guardians, and attorneys. The courts treat a fiduciary's legal duties quite seriously. A breach of fiduciary duties can result in liability for damages, civil contempt and even jail sentences.

CONSTRUCTION RISKS

Contract documents will often designate a project owner as a fiduciary after an insured loss. As a fiduciary, the owner becomes bound by a legal duty to act in the best interest of each "party in interest" for the purpose of safeguarding the funds that were entrusted to him by the insurance companies who were obligated to pay for the insured loss. A practical example illustrates how this works.

Suppose that a fire incinerates the top floor of an office building while it is under construction. The owner wants the fire damage repaired in order to continue with construction. All of the contractors, subcontractors and suppliers want to continue with construction too. Their contracts, in fact, obligate them to finish the work. But they are nervous. The contract documents provide for an excusable delay in the event of a fire but no extra compensation. It's going to take a lot of money to remove and replace all of the fire damage. There is a property insurance policy and it covers fire damage. The problem is that the owner owns the policy. How can all the contractors, subcontractors and suppliers whose work was ruined by the fire be assured of getting the money that they need to repair the damage?

To make the owner a fiduciary is to mitigate their concerns. As a fiduciary, the owner has the legal duty to act in the best interest of each "party in interest" or suffer the consequences for not doing so. The contractor and each subcontractor and supplier affected by the fire is a party in interest. Their interest is served by receiving the money that they need to repair their portion of the fire damage. The fiduciary is obligated to make sure that that happens. If any party in interest is still nervous he can cause the owner to acquire a surety bond to guarantee payment. The owner has to segregate insurance proceeds in a separate account and all parties in interest have to agree with how funds will be distributed. Binding dispute resolution is available for any disputes between the parties in interest.

14.3 Insurance Products

Contractors purchase insurance for claims based on:

- Bodily injury, occupational sickness or disease, or death of the contractor's employees;
- Bodily injury, sickness or disease, or death of any person other than the contractor's employees;
- Bodily injury, death, or property damage due to motor vehicles;
- Bodily injury or property damage arising from completed operations (arising after the contractor has left the site); and

- Personal injury for injuries such as libel, slander, and false arrest as opposed to bodily injury such as physical harm to a person;
- Damage or destruction other than to the work itself (damage to adjacent property);
- Damage to construction equipment and motor vehicles;
- Workers' compensation or similar disability benefit or employee benefit acts; and
- Indemnification.

Contractors ordinarily cover these claims by grouping different types of insurance products; such as commercial general liability insurance, builders risk insurance, equipment floater insurance, automobile insurance, umbrella excess liability insurance, and workers' compensation insurance.

Commercial General Liability Insurance

Commercial general liability insurance [CGL] covers financial losses sustained by third parties that are attributable to construction activities. Third parties are anybody other than the insured party, the insurance company, and named insureds. CGL covers bodily injury, including death, or property damage or both. Three distinctive types of coverage are common: premises and operations; completed operations; and independent contractor. Losses due to errors or omissions in the design of the project are excluded from CGL coverage.

Premises and operations coverage is for claims arising from events occurring during construction involving the building under construction, adjacent buildings, or the construction company's property. Independent contractor coverage is for claims arising from contractual indemnity clauses. Completed operations coverage is for claims arising from events occurring after substantial completion and until such time as the contractor's exposure to liability terminates, usually dictated by the statute of limitations in the location where the work had been performed.

Typical CGL claims include damage to property adjacent to the construction site, injury to visitors on the construction site, and damage to underground pipe and overhead wires. Most CGL policies exclude liability associated with automobiles, aircraft and marine vessels; damage by the contractor's employees to the work in progress or the contractor's property; injury to the contractor's employees; environmental hazard claims; and operations with 50 feet of rail lines.

Construction Risks

CGL policies are either occurrence-based or claims-based. An **occurrence-based policy** insures covered claims regardless of when the claim is made, provided that the insurable event occurred during the policy period. A **claims-based policy** insures covered claims only when the claim is made during the policy period, regardless of when the insurable event occurred. Claims-based policies cost less but expose the contractor to more risk. Occurrence-based policies cost more and protect the contractor from more risk.

Whether or not the policy covers or excludes the cost of defense in a lawsuit is an important consideration. If covered, the amount of money available for claims will be reduced by the cost of defending the claim. If excluded, the cost of defense can be added as an endorsement to the policy, in which case the money available for claims will not be reduced by the cost of defending the claim.

The contract documents limit a contractor's insurance liability to either the amount that is required by law or the amount stated in the contract documents, whichever is greater. The phrase **limits of liability** means the maximum dollar amount the insurance company will pay for any particular liability. Contractors are usually barred from starting any work without insurance coverage. They are required to maintain coverage until at least final completion. Insurance coverage for indemnification and for liabilities that extend beyond final completion must continue in full force and effect until terminated by the statute of limitations in the state for which the work is being performed.

Contractors usually must file **certificates of insurance** with the owner prior to starting the work. A certificate of insurance is a document issued by an insurance company that verifies the purchase of a particular type of insurance, the effective date of the policy, the dollar amounts of coverage for different types of liabilities, and the limits of liability for the policy. Additional certificates must be filed each time a policy is replaced or renewed. Additional certificates may be required for final completion. All certificates must say that a specified number of days, typically 30 days, of prior written notice will be given to the owner prior to cancellation or expiration of any policy. The words **general aggregate** refer to limits of liability for the contractor across all projects insured by the contractor. If the limits of liability for the current project are reduced or exhausted by claims on other projects the contractor is obligated to promptly notify the owner.

The contractor may be required to name the owner, designer and/or designer's consultants as **named insureds** under the contractor's commercial liability insurance policy. This prevents insurance companies from subrogating against other project participants and effectively reduces legal expenses for all named insureds.

The owner may be required to purchase liability insurance too. Most owners will ordinarily have pre-existing liability insurance but they may need to increase the scope or coverage of their existing policies for the additional risks associated with construction activities.

Builders Risk Insurance

Builders risk insurance is a special type of property insurance that indemnifies an "insured party" for damage to the work in progress due to casualties such as explosions and fire, ice and snow, theft and vandalism. For both private and public building projects the owner purchases the builders risk insurance. Because the owner buys the insurance, the owner is the "insured party" despite that it is the contractor's work that is at risk due to casualties. This is common industry practice. In the event of damage to the property under construction the owner administers the proceeds of builders risk insurance, in trust, on behalf of the contractor and all subcontractors and suppliers whose work was affected.

Destruction of the work in progress places choices and attendant duties upon the owner. The owner may choose to terminate the contract and keep the insurance proceeds less any amount owing to the contractor. The owner may also choose to have damaged property repaired by the contractor and then proceed with completing the work. With the later choice, the owner must issue a change order to compensate the contractor for repair work. Insurance proceeds do not limit the amount of this change order. The contractor may want to be relieved from any liability for the loss of use of the owner's property due to fire or other hazards. And the owner may want to purchase loss of use insurance.

Some highway and heavy civil projects go uninsured for casualty, leaving the contractor to purchase builders risk insurance at its option (or to not purchase at its peril). When the contractor purchases builders risk insurance the contractor becomes the "insured party." Destruction of the work in progress by fire is the most expensive component of builders risk insurance premiums, however, and highway and heavy civil projects are not apt to suffer fire damage.

Builders' risk insurance is more expensive than ordinary property insurance because casualties are much more likely to happen during construction. Buildings are especially vulnerable to fire damage during construction because fire prevention mechanisms may be only partially installed or not yet in place.

Builders risk insurance is ordinarily purchased in the amount of the total value of the project on a replacement cost basis. To protect the contractor's interests most contract documents will require the owner to maintain this insurance until final payment has been made or until no person or entity has an insurable interest remaining in the project.

Some contract documents will stipulate that the owner must either insure as required or inform the contractor otherwise in writing. If the owner informs the contractor that it is not insuring the work, the contractor then may purchase builders risk insurance and charge the cost to the owner as a change order. In the event that the owner neither buys property insurance, nor so informs the contractor in writing, the owner may become liable for any losses that would have been covered by the property insurance.

Builders risk insurance can be purchased on an **all risk** basis or as insurance against **named perils**. All risk insurance covers every risk except those that are specifically listed as being excluded. Named perils insurance covers only those risks that are specifically listed and excludes everything else. In practice, both types of policies can end up covering the same risks. Contract documents often stipulate an all risk type of property insurance with coverage for fire (including smoke damage and other consequences of fire), theft, vandalism, malicious mischief, collapse, earthquake, flood, windstorm, falsework, testing and startup, temporary buildings, debris removal, demolition, and reasonable compensation for the designer's and contractor's services required as a result of such insured losses.

The contract documents may require the owner to purchase and maintain **boiler and machinery insurance**. This is a special form of property insurance that covers losses due to the breakdown of any machinery, such as a compressor, that is an integral part of the building under construction. Builders' risk policies ordinarily cover materials stored both on site and off site but not the contractor's equipment. It behooves a contractor to confirm the limitations of coverage.

The owner may be required to submit copies of all property insurance policies to the contractor prior to starting the work. The contractor may be given similar instructions regarding CGL insurance. But where the contractor must submit certificates of CGL insurance, the owner may have to submit the actual property insurance policies. This is so that the contractor can verify that his interests are being served in all conditions, definitions, exclusions and endorsements. Each policy must say that a set number of days, typically 30 days, of prior written notice will be given to the contractor prior to cancellation or expiration of the policy.

The owner may be required, if requested in writing by any party in interest, to acquire a surety bond to guarantee performance of his duty to allocate insurance proceeds. All parties in interest have to either agree on this allocation or resolve their dispute through binding dispute resolution.

Mortgagees, like the construction lender, may also be designated as parties in interest. Contractors are also required to pay all of their subcontractors and suppliers and to require all of their subcontractors to pay their sub-subcontractors. The owner may be authorized, as fiduciary, to settle all losses with insurers unless any party in interest objects in writing within a set number of days, typically five days, of the occurrence of the loss. Objections move settlement to binding dispute resolution.

All project participants may be required to waive their right of subrogation against all other project participants. This effectively bars the builders risk insurer from subrogating a loss. Because the builders risk insurer would ordinarily expect to subrogate, this waiver of subrogation provision must be disclosed before purchasing the builders risk insurance. The rights of parties in interest to insurance proceeds being held in trust by the owner as fiduciary are unaffected by this waiver of subrogation. The waiver of subrogation provisions may be extended to property insurance that the owner might purchase for adjacent property, other property on the site, or purchased later.

Equipment Floater Insurance

Equipment floater insurance covers the contractor's heavy construction equipment like backhoes, excavators, and dozers; temporary structures like concrete forms and falsework; and materials and supplies that will not be incorporated into a construction project. Contractors purchase equipment floater insurance without reference to specific projects. Policies cover equipment, temporary structures, materials and supplies at all times, regardless of whether in storage, in transit, or on the jobsite.

Vehicles that are licensed for highway operation are not covered by an equipment floater but by automobile insurance. Floating marine equipment insurance covers marine equipment. Transportation floater insurance covers physical damage to property other than equipment while it is being moved from place to place. Loss or damage to leased or rented equipment is either excluded or available as an endorsement at extra cost. Contractors typically add endorsements to insure against failure due to maintenance and repair, and damage due to equipment overload especially crane boom overloads.

Automobile Insurance

Automobile insurance is for contractor's vehicles that are licensed to operate on public highways. Automobile insurance covers physical damage to the contractor's vehicles. It also covers damage to the property of others and liability to others for bodily injury and death.

Umbrella Excess Liability Insurance

Contractors purchase umbrella excess liability insurance in order to increase their aggregate CGL insurance limits.

The words "aggregate CGL insurance" need explanation. Contractors are required to buy CGL insurance for every job. Ordinarily many such individual insurance policies will be purchased from the same insurance company. Each insurance company will set limits of liability for every job and limits of liability for all jobs, in aggregate, of a particular contractor. A contractor, for example, may have three jobs going, each with $50 million dollar limits of liability, and an aggregate limit of $100 million. If $50 million dollar claims are levied against each of two jobs, the third job would have no coverage left because the aggregate limit would have been exceeded. Umbrella excess liability insurance provides coverage in excess of the aggregate CGL insurance limits of liability.

Workers' Compensation Insurance

Even though the Occupational Safety and Health Act of 1970 [OSH Act] is in place to prevent injuries in order to make workplaces safer, workplace injuries still happen (Occupational Safety and Health Act of 1970 [OSH Act], 1970). Workers' compensation statutes are in place to deal with the consequences of workplace injuries. Workers' compensation statues require almost all employers, whether in the private or public sector, to provide coverage for their full-time and part-time employees.

Because people who perform work as independent contractors are excluded, firms using subcontractors should ensure that those subcontractors are providing workers' compensation for their employees. But even if an employer does not pay workers' compensation premiums for its contract employee (or temps), it is still obligated under the Occupational Safety and Health Administration [OSHA] rules to provide a safe workplace for those workers.

By law, employers must arrange workers' compensation coverage for their employees. They do so by contributing to state workers' compensation funds, purchasing coverage from private insurers, or self-insuring. The available funding options vary in every state. Workers' compensation insurance covers all injuries, illnesses, and deaths that "arise out of and in the course of" employment. The insurance provides employees with partial replacement income, medical care, and rehabilitation.

Workers' compensation insurance is strict liability insurance. Commonly described as "no-fault" insurance, injured workers need not prove that they were blameless for their injury or that anybody else was culpable in order to be compensated. But in exchange for being absolved of blame, employees have no opportunity to sue their employer for their injuries and the dollar amounts of claims are limited by statute. Long term disabilities and death, however, involve processes for which the injured or their families typically secure legal counsel.

The premiums that a construction company pays to workers' compensation insurance providers are determined by a **base rate** and an experience modification ratio. The base rate is determined by an employee work classification that reflects the relative hazard inherent in the trade that is being insured. The **experience modification ratio (EMR)** is based on the contractor's recent claims history. Experienced contractors with good safety records will generally have EMRs below 1.0.

Workers' compensation statutes do not necessarily bar injured workers from suing owners, engineer, architects, other contractors and subcontractors not employing them but active on the job at the time that they were injured. When serious jobsite injuries occur there is a high likelihood that lawsuits will emerge that involve many of the project participants. Generally, contractors are required to indemnify the other participants from liability arising from these types of lawsuits. Insurance coverage is available for these claims as endorsements – and not inexpensive endorsements - to workers' compensation policies.

Chapter 15 SURETY BONDS

15.1 Introduction

The owner's principal reason for having a written construction contract is that it memorializes its contractor's promise to perform the work of the contract. On a material breach of that promise the owner is empowered with remedies at law. That is all well and good except for the shortcomings inherent in remedies at law. A very long time can elapse between the breach and its remedy. During that time the owner must commit a substantial amount of resources to the dispute resolution process. The work of the contract still needs to be completed after a contractor is terminated for material breach and any replacement contractor that is hired will have to be paid by the owner, usually at premium rates. The owner only hopes to recover those payments from the breaching contractor. Yet whatever amounts are recovered will be reduced by attorneys' fees and other litigation or ADR expenses. The worst possible outcome is that the remedy might be nothing more than a useless, uncollectable court judgment for damages against a now bankrupt contractor…and a still uncompleted project.

In order to protect their interests, owners will want a surety standing by; ready to step in and guarantee completion of the work should the contractor falter. Lending institutions may compel an owner to obtain surety bonds in order to protect their construction-lending capital. A **surety** is an independent third-party with ample financial resources who has the capability to discharge the duties of the contractor either by completing the work of the contract or by paying the owner to have it completed. A surety's guarantee is expressed in a written, tripartite agreement between the owner, the contractor and the surety, known formally as a **surety bond** and informally as a **bond**.

The surety is the **guarantor** of a surety bond. The guarantor ensures completion of the work of the contract. The owner is the **obligee** (pronounced äb-le-'jē) of a surety bond. The surety bond protects the obligee.

The contractor is the **principal** of a surety bond. The principal purchases the surety bond and indemnifies the surety for any costs incurred by the surety in executing its guarantee. A general contractor may require its key subcontractors to obtain surety bonds in order to mitigate the effects of subcontractor defaults. When a subcontractor purchases a surety bond the general contractor is the obligee and the subcontractor is the principal.

Bonds are not unlimited guarantees. Bonds are issued with a **face value**, sometimes called a **penal sum**. A bond's face value sets a cap on the monetary amount of a surety's liability under a bond.

From the perspective of the obligee, a surety bond acts like insurance. It protects the obligee from harm should the principal fail to perform its obligations. From the perspective of the principal, however, a surety bond is nothing like insurance. A surety bond exposes a principal to liability. An indemnification agreement will be formed between the guarantor and the principal prior to executing any surety bonds. In this indemnification agreement the principal pledges its business and personal assets to reimburse any costs incurred by the guarantor in discharging the principal's duties.

A contractor purchases bonds from its surety and passes their cost onto the owner. Owners must convey their request for surety bonds in their bid documents in order for contractors to price them into their bids. A contractor will not actually purchase the bonds until receiving notification of award at which point the contractor must facilitate the execution of all required surety agreements prior to commencing work on the contract.

15.2 The Miller Act

The Miller Act is a federal statute that mandates **performance bonds** and **payment bonds** for all federal construction contracts in excess of $100,000 (The Miller Act of 1935 [Miller Act], 1935). Many of the states have enacted similar statutes, known as Little Miller Acts, requiring performance bonds and payment bonds for public projects in their state.

Performance Bonds

A **performance bond** is a guarantee by a surety of the performance of a particular contract. Miller Act performance bonds must be written for a face value of 100% of the contract amount. Private owners will typically request 100% performance bonds as well.

For the surety to step in and discharge the contractor's performance the contractor must be in material breach of contract. The surety is not obligated to intervene for a minor breach or for a material breach by the owner. The breach must be material and the contractor must be culpable.

A surety's discharge of a contractor's performance does not relieve the contractor of any liabilities. The surety and its contractor are jointly liable pursuant to all duties under the owner-contractor agreement. This joint liability infers that the surety cannot have any greater liability than the contractor would have had with respect to the owner. The surety also jointly acquires all defenses to liability that the contractor possesses.

The surety may have additional defenses, known as **technical defenses**, arising from the language of its surety agreement. Change orders may precipitate a technical defense. The surety bond only requires the surety to guarantee the performance that had been stipulated in the contract documents at the time that the owner-contractor agreement was signed. Change orders and other modifications to the contract documents need the consent of the surety. If the surety had not been informed of a change or modification, or had been informed but had not consented, the surety may be relieved of its obligations, if not in whole then at least to the extent of the change or modification.

The surety may meet its obligation to complete performance in various ways:

1. Take over completion of the work;
2. Buy out its obligation to the owner with a cash settlement;
3. Hire a replacement contractor to complete the work and pay any extra costs thereto; or
4. Cause the contractor (the principal) to cure its default, complete the work of the contract, and pay any extra costs thereto.

Particularly when the contractor (the principal) had provided little, if any, self-performed work, a surety with adequate capabilities may wish to take over completion of the work (item 1 above). A takeover requires a negotiation between owner and surety over terms of the takeover. This negotiation results in a **takeover agreement**. All subcontracts are usually assigned to the surety who then interjects the requisite construction management skills and financial resources to bring the project to completion.

It is prudent for private owners to consult with legal counsel and insurance advisors whenever a performance bond is contemplated.

Payment Bonds

A **payment bond** is a guarantee by the surety that laborers, tradespersons, subcontractors and suppliers, will be paid for work performed under a construction contract. Typically, the payment bond will have the same face value as the project's performance bond. A payment bond differs from the performance bond in that the owner, or other obligee, cannot make a claim under a payment bond. Only laborers, tradespersons subcontractors or suppliers seeking payment may make claims.

A payment bond is an alternative to a mechanics' lien for securing payment for performing work. Payment bonds are particularly useful in public construction where liens cannot be levied on public property. Despite that they serve the same purposes, subcontractors and suppliers will sometimes avail themselves of both mechanisms on private construction projects: filing mechanics' liens while making claims under a payment bond at the same time.

Payment bond claimants are classified into tiers. Laborers, tradespersons, subcontractors and suppliers of the general contractor (or prime contractor(s)) are the 1st tier. Laborers, tradespersons, subcontractors and suppliers of the 1st tier subcontractors and suppliers are the 2nd tier. Laborers, tradespersons, subcontractors and suppliers of the 2nd tier subcontractors and suppliers are the 3rd tier, and so forth.

Statutory law and public project regulations will often limit the rights of payment bond claimants with respect to the tier that they are in. Payment bond protection under the federal Miller Act, for example, is extended only to the first two tiers. And general contractors are not allowed to insert empty shell subcontractors into the project hierarchy just to push those in the lower tiers out of the range of protection under their payment bond.

Whether one is a classified as a subcontractor or a supplier is also a crucial distinction under the Miller Act. The Act will protect suppliers of the general contractor (or prime contractor(s)) and suppliers of subcontractors but it will not protect suppliers of suppliers. Although the Miller Act is silent on how to distinguish a subcontractor from a supplier, case law has provided practical definitions. Any claimant who employs laborers and/or tradesmen to perform work onsite will usually be classified as a subcontractor. Claimants who only provide materials or equipment will usually be classified as suppliers. A claimant who fabricates materials and systems offsite might also be classified as a subcontractor if the materials and systems that they fabricated were unique to the project or in the event that they would have little commercial value were they not used for the project.

The surety's technical defenses for payment bond liability usually arise from deficient notices. The Miller Act requires subcontractors and suppliers in the 2nd tier to notify the general contractor or prime contractor within 90 days of the date that they last performed any work that is the subject of a payment bond claim. Notification should be in writing so that its delivery can be verified; it must alert the general contractor or prime contractor that a claim is being made under the payment bond; it must state the amount being claimed; and it must state what labor was provided or to whom the materials were delivered to. Miller Act notice requirements do not apply to 1st tier contractors.

Payment bond protection under the Miller Act extends to labor and material that is incorporated into the work of the contract. This ordinarily includes all materials that are delivered to the jobsite, even materials that were damaged in transit making them unsuitable for incorporation into the work. Materials that are consumed or made useless, such as welding rods, formwork, falsework and temporary structures, incidental work, warranty work, equipment rentals, operating leases for construction equipment, transportation and delivery costs, union contributions, and contributions to employee benefits programs, are all considered incorporated into the work of the contract. Under appropriate circumstances food and lodging expenses are included. Change orders are considered incorporated into the work but compensable delays are generally not within the scope of a payment bond.

Lawsuits to enforce Miller Act payment bonds must be filed within one-year of the day that the work that is the subject of the payment bond claim was last performed. Punch list work can initiate this one-year tolling period. Warranty work performed after final completion is out of the scope of payment bond protection. In no event can a lawsuit be filed more than one year after final completion.

15.3 Other Surety Bonds

Bid Bond

A **bid bond** is a guarantee to the owner, by a surety, that the surety will assume liability if a bidder, upon notification of award, fails to execute the owner-contractor agreement and furnish all surety bonds and insurance policies that are stipulated in the contract documents. Bid bonds are purchased by bidders, as principals of the bond, and submitted with their bids.

Bid guarantees are typically required with competitive sealed bids for public construction contracts. A bid guarantee may consist of a postal money order, certified check, cashier's check, irrevocable letter of credit, certain

U.S. Treasury bonds and notes, or a bid bond. They are typically in the amount of 10% of the contract value. Bidders who do not furnish a bid guarantee with their bid, in the proper form and amount, may have their bid rejected as nonresponsive (FAR §52.228-1, 1984).

The contracting officer returns bid guarantees, other than bid bonds, to unsuccessful bidders, and to the successful bidder when the owner-contractor agreement is signed and all performance bonds, payment bonds, other surety bonds, and certificates of insurance are furnished.

Bidders who elect to supply a bid bond as a bid guarantee are not reimbursed for their expenses when they do not win the award. Bid bond expenses are a cost of doing business. Fortunately, sureties rarely charge for bid bonds. A surety will investigate a contractor's capabilities and if willing to issue performance and payment bonds they will usually offer to issue a bid bond free of charge in anticipation of selling performance bonds and payment bonds when and if the contractor wins the award.

Ordinarily, damages for failure to execute the agreement and furnish all bonds and certificates of insurance are limited to actual damages. Actual damages are calculated as the difference between the defaulting principal's bid and the next lowest responsible and responsive bid. For bid guarantees other than bid bonds, actual damages are deducted from the guarantee amount and the remaining balance, if any, is returned to the defaulting bidder.

If the bid guarantee is a bid bond, damages are assessed against the surety. In some jurisdictions damages are statutorily set at the penal sum of the bond. This type of bid bond is known as a **forfeiture-type bond**. Owners may recover the full penal sum of a forfeiture-type bond regardless of the amount of actual damages that they incurred. In other jurisdictions, an owner's recovery under a bid bond is limited to its actual damages. This type of bid bond is known as a **damages-type bond**.

The surety may have technical defenses that arise from failure to inform the surety of addenda or other changes, informing the surety of addenda or other changes where the surety had not consented, unjustified delay in notification of award, and substantial and apparent bid mistakes.

License Bond

A **license bond** is a surety bond, the purpose of which is to protect consumers from defective construction that arises from contractor licensing law violations. The license bond guarantees that individuals granted a license will meet the obligations under that license.

In California, for example, all licensed contractors are required to post security with the State. Performance of any construction work in excess of $500 in contract value requires a current contractor's license and a license bond. Contractors must post a license bond in the amount of $12,500, "…for the benefit of consumers who may be damaged as a result of defective construction or other license law violations and for the benefit of employees who have not been paid wages that are due to them, as a requirement for the issuance of an active license, reactivation, and maintenance of actively renewed license" (California Business & Professions Code [Cal B&P Code] §7071.6, 2014). Cash or a certificate of deposit in the full amount of the surety bond may be deposited in lieu of the license bond.

Consumers, suppliers, or an employee of a contractor may file a claim against the bond for any acts of the licensed contractor in violation of the State of California contractor licensing laws. The surety will endeavor to validate each claim. In the event that the contractor is found culpable of a contractor licensing law violation, the surety incurs liability to pay damages to the bond claimant.

Lien Release Bond

A **lien release bond** is a surety bond that is utilized to discharge a mechanics' lien. Because a mechanics' lien goes with the property, a new owner acquires the liability attendant to a mechanics' lien along with transfer of ownership of real property. Consequently, real property will rarely be sold with a lien attached to it. A lien release bond enables the sale of real property.

The property owner is usually the principal for the bond. The beneficiary of the bond is the party who had placed the mechanics' lien on the property. The surety guarantees payment to the beneficiary if the beneficiary's claim is found to be legitimate.

Because lien release bonds are considered risky, the principal may be required to post collateral or other assets to secure the bond. Bonds are generally filed with the county recorder where the work is being performed and the bond cannot be cancelled until the principal provides documentary evidence that the liability under the lien claim has been terminated.

In California, lien release bonds must be in the amount of 125% of the value of the mechanics' lien that the bond discharges.

15.4 Surety Transactions

A contractor's fitness to obtain a surety bond is a prequalification for most public contracts and for many private contracts as well. Sureties will determine a contractor's fitness by evaluating the contractor's capital structure, capacity to do work, and strength of character. The aggregate value of the bonds that a surety is willing to issue on behalf of a contractor serves to delimit the revenues that the contractor can aspire to.

A surety does more than sell bonds. Sureties are themselves competent business managers. Contractors who are experiencing difficulties will find that sureties can help them find sources of financing and otherwise help them to manage their projects to successful completions. Prequalification by a surety increases the number of projects for which a contactor can qualify; payment bonds ensure protection for subcontractors, suppliers and employees; and the surety can provide technical, managerial or financial assistance to a contractor. A surety's prequalification provides owners with a pool of qualified bidders; payment bonds reduce the owner's risk of liens and financial losses; and performance bonds increase the likelihood of timely project completion and protect against defective materials and workmanship.

Surety Investigations

A surety's business is to sell bonds. Yet they face financial liabilities when a contractor defaults on its performance obligations. Surety bonds are a calculated risk. In order to properly manage that risk, sureties will carefully investigate a contractor's business operations before offering to sell bonds to them. And when they do sell they set firm limits on the aggregate number and size of their bond issues.

Default happens to contractors despite their best intentions. A default may bring to light deficiencies in a contractor's business operations. Typical operational deficiencies include hyper growth, imprudent diversification, inadequate financial controls, and deficient project management. A surety's investigation provides an in-depth look at the contractor's entire business operation. Its purpose is to determine the contractor's ability to meet its current and future contractual and financial obligations.

A surety investigation occurs in two stages. The 1st stage surety investigation is a prequalification test: to determine if the surety is willing to offer bonds to a contractor and if so what limits would be placed on the type, aggregate number and size of its bond issues. The goal of the 2nd stage surety investigation is to determine if they are willing to offer bonds for a particular project to a contractor who had prequalified in their 1st stage investigation.

The focus of a 1st stage surety investigation is on the principals and senior-level managers of the construction firm. The investigation explores their professional ability, reputation, integrity, track record in particular industry segments, the strength of their financial reports, and their access to capital. Separate lines of inquiry explore character, capacity, and capital. Character inquiries look into the firm's reputation for fair and ethical business dealings with owner, subcontractors, suppliers, and creditors. A worthy contractor should have a good reputation within the construction industry. Capacity inquiries look into the firm's history, its organization and reporting structure, the professional ability of its key personal, its track record on past projects, its performance on current projects, its inventory of tools and equipment, and its trade references. Capital inquiries examine the firm's financial statements, its management accounting and controls, its lines of credit, assets, proforma financial forecasts, ratio and trend analysis, and risk assessment.

Topics of a 2nd stage surety investigation include the contractor's current work backlog, recent bidding record, present working capital, project milestone progress, terms of payment and cash flow, previous experience with the particular scope of the project for which surety bonds are sought, and the value of the work to be bonded.

Surety bonds are required for public and private projects alike. Private project owners might not bother with prequalifying their bidders, relying instead on a surety's investigation: reasoning that any contractor who prequalifies with a surety is good enough for their project. To many contractors, a prequalification by a surety is the difference between doing business and going out of business.

But even a prequalified contractor needs a prudent bidding strategy. A contractor who bids a lot of projects just might win a lot of awards. Its surety, however, will be reluctant to offer bonds when they start to exceed the contractor's capacity to perform work. Suddenly unable to purchase more surety bonds, a contractor could default on its bid bond. Sureties are not legally obligated to offer performance and payment bonds after issuing a bid bond but they usually do. For that reason, sureties evaluate each request for a bid bond and they will decline to offer them whenever the contractor is believed to have an excessive number of active bids.

Indemnification Agreements

After a contractor has been qualified but before any bonds are offered, the contractor must enter into an indemnification agreement with its surety. This **indemnification agreement** exchanges the surety's commitment to offer surety bonds for the contractor's pledge to indemnify the surety for any losses that the surety might sustain due to a default by the contractor.

In all but the largest, publicly traded construction companies, the contractor's pledge requires personal indemnity from the principals of the contractor's firm; its proprietor, partners, members, shareholders, officers or directors. Personal indemnity obligates the firm's principals with joint and several liabilities for actual losses sustained by the surety, even if these losses are in excess of the project amount. **Joint and several** liabilities enable liability to be apportioned to one or more parties or to the entire group of principals, at the discretion of the surety. Both business assets and personal assets may have to be pledged to support an indemnification agreement.

Bond Premiums

The purchase price of a surety bond is known as its **bond premium**. Bond premiums are passed through to the owner, usually as a line item in the contractor's bid. Bond premium rates are based upon project classification. There are four classifications: A-1, A, B, and Miscellaneous. Most projects fall under classification B, general construction work. If the project has multiple classifications, the rate is based upon the highest classification. Good standing with the surety yields lower bond premium rates.

Bond premiums are calculated as a bond premium rate times the contract value. Bond premium rates range from 0.5 to 1 % for excellent contractors; 1 to 2% for good contractors; and up to 10% for marginal contractors. Bond premium rates are sometimes assessed on a sliding scale relative to the size of the contract.

PART III - CONSTRUCTION REGULATIONS

Chapter 16 LICENSING AND ENTITLEMENT

16.1 Introduction

The goal of contractor licensing is to protect the public from defective construction. To that end, the licensing and regulation of contractors is administered at the state level by contractor state licensing boards. Each state has developed different approaches to meeting their responsibility to protect the public through licensing and regulation. Typical tactics include enforcing minimum competencies in a particular trade, establishing standards for representation of honesty, demonstrating adequate financial capacity to conduct contracting work, and verifying proficiency and knowledge of construction laws and regulations. There is a common regulatory emphasis on experience in the various trades.

The goal of entitlements is to implement congressional preferences for small businesses, affirmative action mandates, and domestic materials purchases.

16.2 Contractor Licensing

Licensing requirements differ among the various States:
- Residential and commercial contractors must be licensed in fifteen States;
- Mississippi licenses commercial contractors only;
- Michigan and Minnesota license residential contractors only;
- General contractors are licensed in eight States;
- Roughly half of the States license plumbing and electrical contractors.
- A variety of specialty contractors, including lead and asbestos abatement contractors are licensed among the States

CONSTRUCTION REGULATIONS

	Commercial Contractor	Residential Contractor	Home Improvement	General Contractor	Business Registration	Electrical	Plumbing	Abatement Contractor	Other Specialty Trades	City or County License
Ten States	10.	10.								
California	✓	✓		✓		✓	✓	✓	✓	
Colorado						✓	✓	✓	✓	✓
Connecticut					✓					
District of Columbia					✓				✓	
Florida				✓		✓	✓		✓	
Georgia	✓	✓				✓	✓		✓	
Hawaii				✓		✓	✓		✓	
Idaho	13.	13.				✓			✓	
Illinois							✓		11.	✓
Indiana							✓			✓
Iowa					✓					✓
Kansas					✓					
Kentucky							✓		✓	
Louisiana	✓	✓				✓		✓	✓	
Maine			1.			✓	✓			
Maryland		✓				✓	✓			
Massachusetts		✓				✓	✓			
Michigan		✓				✓	✓			
Minnesota		✓							✓	
Mississippi	8.									✓
Missouri					✓					✓
Montana						✓	✓	✓	✓	
Nebraska						✓	✓			✓
New Hampshire						✓	✓	✓	✓	
New Jersey			2.			✓				
New York					✓			✓	✓	✓
North Carolina				6.		✓	✓			
North Dakota				7.		✓	✓			
Ohio										✓
Oklahoma				✓		5.	5.		5.	
Pennsylvania										
Rhode Island			3.			✓	✓		✓	
South Carolina	9.	9.								
South Dakota						✓	✓	✓	✓	
Texas						✓	✓			✓
Vermont						✓	✓	✓	✓	
Virginia						✓	✓	✓	✓	
Washington					✓	✓	✓			
West Virginia				4.		✓	✓	✓	✓	
Wisconsin				12.						
Wyoming						✓				✓

1. State registration for home repair services sales.
2. State registration for new homebuilders.
3. State contractor's license for work on one to four-family dwellings
4. State license for all contractors.
5. State license for resident electrical, mechanical and plumbing contractors.
6. State contractor's license for all jobs over $30,000
7. State contractor's license for all jobs over $2,000
8. State contractor's license for commercial work over $100,000, city or county over $50,000
9. State contractor's license for commercial work over $5,000 and residential over $200
10. Alabama, Alaska, Arizona, Arkansas, Delaware, Nevada, New Mexico, Oregon, Tennessee & Utah
11. State license required for roofing work.
12. State credentialing required.
13. State licensing for public works projects

Figure 16.1 State-By-State Contractor Licensing at a Glance

- Special licensing and registration requirements for home improvement contractors are enacted in five states; and
- Many states delegate licensing to city or county governments. (Contractor's Licensing Service, Inc. [CLSI], 2014; Circo & Little, 2009).

Figure 16.1 summarizes contractor-licensing requirements in the various states.

There is an evolving trend in the licensing of construction managers. The states of California, Idaho and Oklahoma license construction managers as separate professions, but only for the construction management of public works projects (Sweet, Schneier & Wentz, p. 159, 2015). A growing number of states regulate construction managers as contractors, engineers or architects (Kolton & Montgomery, 2007). Most of these states regulate them as either general contractors or design professionals but some regulate them as a hybrid of both (Kolton & Montgomery, 2007). Other states impose no specific regulations on construction managers.

In those states that require contractor licensing, the desired conduct by contractors is motivated by two primary means. The first means leverages power vested in the contractors state licensing boards. These boards are authorized to issue licenses and to adjudicate licensing violations. Suspensions, revocations and other disciplinary actions can be levied for repeated incidences of patent or latent defects or omissions. A license revocation is a severe remedy because it removes a contractor's legal ability to sustain a contracting business. Some statutes mandate criminal sanctions for certain licensing law violations. Incidents where the courts have ordered criminal punishment have been rare, however.

The second means of motivating contractor conduct is through legislation that bars unlicensed contractors from enforcing their construction contracts in court. This leaves unlicensed contractors without a legal remedy in law for damages. The courts of the various states have been mixed in their support of such legislation.

Contractor Licensing in California

The statutory authority for licensing and regulating contractors in California is Business and Professions Code §§7000 *et seq.* (Cal B&P Code, 2014). Regulatory authority is vested in the **Contractor's State License Board [CSLB]** organized under the California Department of Consumer Affairs. Duties of the CSLB include: licensing exam preparation and administration; issuing licenses and maintaining licensing records; investigating complaints against contractors; issuing citations, suspensions and revoca-

CONSTRUCTION REGULATIONS

tions of licenses; and seeking criminal and civil sanctions against licensing law violators (Contractors State License Board [CSLB], n.d.).

California contractors must be licensed to perform any construction work with a value of $500 or greater. A contractor's license is issued for a two-year period and is renewable every two years. Licenses are issued in three general licensing categories and 43 separate specialty contractor trades (See Figure 16.2).

(A) General Engineering Contractor
(B) General Building Contractor
(C) Specialty Contractor
    C-2 - Insulation and Acoustical Contractor
    C-4 - Boiler, Hot Water Heating and Steam Fitting Contractor
    C-5 - Framing and Rough Carpentry Contractor
    C-6 - Cabinet, Millwork and Finish Carpentry Contractor
    C-7 - Low Voltage Systems Contractor
    C-8 - Concrete Contractor
    C-9 - Drywall Contractor
    C10 - Electrical Contractor
    C11 - Elevator Contractor
    C12 - Earthwork and Paving Contractors
    C13 - Fencing Contractor
    C15 - Flooring and Floor Covering Contractors
    C16 - Fire Protection Contractor
    C17 - Glazing Contractor
    C20 - Warm-Air Heating, Ventilating and Air-Conditioning Contractor
    C21 - Building Moving/Demolition Contractor
    C23 - Ornamental Metal Contractor
    C27 - Landscaping Contractor
    C28 - Lock and Security Equipment Contractor
    C29 - Masonry Contractor

C31 - Construction Zone Traffic Control Contractor
C32 - Parking and Highway Improvement Contractor
C33 - Painting and Decorating Contractor
C34 - Pipeline Contractor
C35 - Lathing and Plastering Contractor
C36 - Plumbing Contractor
C38 - Refrigeration Contractor
C39 - Roofing Contractor
C42 - Sanitation System Contractor
C43 - Sheet Metal Contractor
C45 - Sign Contractor
C46 - Solar Contractor
C47 - General Manufactured Housing Contractor
C50 - Reinforcing Steel Contractor
C51 - Structural Steel Contractor
C53 - Swimming Pool Contractor
C54 - Ceramic and Mosaic Tile Contractor
C55 - Water Conditioning Contractor
C57 - Water Well Drilling Contractor
C60 - Welding Contractor
C61 - Limited Specialty
ASB - Asbestos Certification
HAZ - Hazardous Substance Removal

Figure 16.2 California Contractor Licensing Classifications

The licensing exam is a full-day, written test. Half of the exam is focused on construction law and business topics while the other half is focused upon trade practices. Upon passing the exam, an applicant becomes a qualifying individual. A **qualifying individual**, or simply **qualifier**, is the person listed on the CSLB records who meets the experience and examination requirements for a license. A qualifying individual is required for every classification on each license issued by the CSLB.

To become a qualifier, eligible applicants must identify the licensing category that they are applying for, pay a $250 application fee, and sit for a licensing examination. To be eligible to sit for the examination, an applicant must be at least 18 years old and he or she must provide evidence of working experience at the journeyman, foreman, or supervisor level in the licensing classification being applied for, or as a contractor at the classification being applied for, for a total of four years out of the ten previous years. Written documentation certifying technical training, apprenticeships or education may be submitted as a substitute for experience. Up to three years of education credit may be granted.

Licenses are issued to contractor business entities that have any of the following types of business structures: sole proprietorship (an individual owner), partnership, joint venture, limited liability company (LLC) or a corporation. A license is not normally considered an asset owned by the business entity, however. While a license is issued to the business entity, each license is also linked to an individual who is a qualifier for that business entity in the same licensing category or categories as the business entity's license. The qualifying individual does not necessarily have to remain the same over time, although a qualifying individual must be in place in order for the license to be valid.

The individual who is linked to a license is designated the qualifying individual for the licensee. The qualifying individual for a licensee must be responsible for the licensee's construction operations. If the qualifying individual is a corporate officer, he or she is known as the **responsible managing officer [RMO]** for the corporation. Any licensee may choose to hire a licensed individual as their qualifying individual. That employee is known as its **responsible managing employee [RME]**.

If a sole proprietorship has a license its qualifier may be either the proprietor or an RME. If a licensee has a partnership license, its qualifier may either be one of the general partners, designated as qualifying partner, or an RME. If a licensee has a corporate license, its qualifier may be either one of the officers listed on the CSLB's records for its license, who is designated as RMO, or an RME.

If a licensee's qualifier is an RME, he or she must be a bona fide employee of the licensee and may not be the qualifier on any other active license. This means that the RME must be regularly employed by the licensee and actively involved in the operation of the business at least 32 hours per week or 80 percent of the total business operating hours per week, whichever is less.

A person may act as a qualifying individual for more than one active license under certain conditions, but no person may serve as the qualifying individual for more than three licensees in any one-year period. If a qualifier disassociates from the third licensee, he or she must wait one year before associating with a new third licensee. An RME can only act as a qualifying individual for one active license at a time.

Active licensees are entitled to contract for work in the classifications that appear on their license. While the license is active, the licensee must maintain a workers' compensation insurance coverage (unless exempted under certain circumstances) and a current license bond. A license bond called a **contractor's bond** is required when a firms' qualifier is a qualified individual or an RMO who owns 10% of more of the voting stock in the corporation. A license bond known as a **bond of qualifying individual** is required if a firm's qualifier is an RME or an RMO who does not own at least 10% of the voting stock of the corporation. If a license has more than one RME or RMO qualifying the license, each qualifier must comply with the qualifier bonding requirements.

If a license is inactive, that is, currently renewed but on inactive status, the holder may not bid or contract for work. Neither the contractor's bond nor the bond of qualifying individual are required for an inactive license. Also, a licensee does not need to have either the proof or exemption for workers' compensation insurance coverage on file with the CSLB while the license is inactive.

Enforcement

A Statewide Investigative Fraud Team [SWIFT] conducts sweep and undercover sting operations in an effort to eradicate unlicensed contractors. Contracting without a license exposes a contractor to CSLB fines of $200 to $15,000 and it is usually a misdemeanor with a potential sentence of up to six months in jail or a $500 fine or both. Repeat violators are subject to increased criminal penalties. It is a felony to contract without a license for any project that is covered by a state of emergency or disaster declaration proclaimed by the Governor of California or the president of the United States. A state prison term can result from any felony conviction.

The California Business & Professions Code requires every licensed contractor who acting as a prime contractor offers a contract for construction, to include the following statement in at least 10-point type on all written contracts:

> "Contractors are required by law to be licensed and regulated by the Contractors' State License Board which has jurisdiction to investigate complaints against contractors if a complaint regarding a patent act or omission is filed within four years of the date of the alleged violation. A complaint regarding a latent act or omission pertaining to structural

defects must be filed within 10 years of the date of the alleged violation. Any questions concerning a contractor may be referred to the Registrar, Contractors' State license Board. P.O. Box 26000, Sacramento, CA 95826" (Cal B&P Code, § 7159.10, 2012).

16.3 Statutory Proscriptions to Contract

In most of the states, contractors can avail themselves of the court system to enforce their contracts. They can enforce their contracts in court, even if they are unlicensed and in violation of licensing statutes that subject them to civil and/or criminal penalties.

Some of the states marshal additional leverage to motivate the conduct of contractors. They do that by prohibiting unlicensed contractors from enforcing their contracts in court. This leaves unlicensed contractors without a legal remedy in law for damages. In its effect, an owner can refuse to pay its unlicensed contractor and the contractor can do little about that.

The states vary in their treatment of contractors:

- The states of California, Florida, Georgia, North Carolina, South Carolina, and Tennessee require a contractor to be licensed in the appropriate licensing category at all times during performance of the work;
- The states of Alaska, Arizona, Hawaii, Oregon and Utah require a contractor to be licensed in the appropriate licensing category continuously from the time of soliciting for work through final completion of the work;
- Contractors do not have to be licensed in Washington (i.e. no experience, education, or examination requirements) but they must register as contractors. Unregistered contractors are barred from enforcing their contracts;
- The States of Michigan and North Carolina will not enforce contracts of unlicensed contractors but will allow them to defend themselves against a breach of contract suit on its merits;
- In Tennessee unlicensed contractors may enforce contracts but only to recover costs, not profit;

CONSTRUCTION REGULATIONS

	Contract Enforcement Denied. Valid license to bid	Contract Enforcement Denied. Valid license to build	Lien rights denied by statute.	*Quantum meruit* barred by statute
Alaska	✓			
Arizona	✓			
California		✓		
Florida		✓		
Georgia		✓	✓	
Hawaii	✓			✓
Idaho			5.	
Michigan		1.		
Minnesota			2.	
North Carolina		3.		
Oregon	✓		✓	
South Carolina		✓		
Tennessee		4.		
Utah	✓		✓	
Washington		5.		

1. Unlicensed residential builder, tradesman, or maintenance and alteration contractor may not sue to enforce contract, but may defend breach of contract suit on its merits.
2. Unlicensed residential contractors are denied lien rights.
3. Unlicensed contractors cannot sue to enforce contract but may defend breach of contract suit on its merits.
4. Unlicensed contractor may enforce contracts but only to recover costs, not profit.
5. Proscription against unregistered contractors.

Figure 16.3 Contract Enforcement Matrix

- The States of Georgia, Idaho, Minnesota, Oregon and Utah apply additional pressure on unlicensed (or unregistered) contractors by denying them the right to file mechanics' liens; and
- In all other states: either contractors need not be licensed or unlicensed contractors have the right to enforce their contracts in court (See Figure 16.3).

The alternative for any contractor who lacks an enforceable contract is arguably a lawsuit in restitution, on principles of unjust enrichment. The courts, however, have largely denied unjust enrichment claims to unlicensed contractors, reasoning that the goal of protecting the public militates against any encouragement of unlicensed contractors, even if this results in financial inequity. The State of Hawaii even goes as far as to invoke a statutory prohibition against *quantum meruit* recovery by unlicensed contractors.

Although some courts will deny restitution to unlicensed contractors, other courts are more sympathetic to the contractor, allowing restitution in certain circumstances: such as with an owner-contractor agreement where the owner knew that the contractor was unlicensed and was using that knowledge in bad faith to avoid paying (*Jackson v. Davis*, 1978). The same results can obtain with contractor-subcontractor agreements where the parties are professional constructors who are fully knowledgeable about licensing restrictions and one is using that knowledge to its unfair advantage (*Enlow v. Higgerson*, 1960).

Unlicensed California Contractors

In California, a contractor cannot enforce its contract in court unless the contractor holds a valid license, in the appropriate licensing category, at all times during performance of its work (Cal B&P Code, §7031(a), 2014). Nor can the innocent party enforce its contract against an unlicensed contractor. The California Courts have rarely allowed damages to be recovered by either unlicensed contractors or the aggrieved but innocent other party in either breach of contract or in restitution. With few exceptions, neither party can bring or maintain an action in court.

However, an aggrieved party who had used an unlicensed contractor can bring an action in court to recover all compensation that had been paid to that unlicensed contractor (Cal B&P Code, §7031(b), 2014). This provision has been enforced even as to a party acting in bad faith, knowing that its other contracting party was unlicensed.

The California courts will apply the substantial compliance exception to mitigate §§7031(a) and (b). **Substantial compliance** applies when the work of the contract was correctly performed and the contractor's lack of a license was the result of a bureaucratic glitch. Substantial compliance may apply in California when the contractor:

1. Had previously been licensed in California;
2. Acted in good faith to maintain its license;
3. Did not or could not in any reasonable way know that it was not licensed when performance on its contract began; and
4. Acted in good faith to timely reinstate its license upon learning it was invalid (Circo & Little, 2009, p. 107).

Substantial compliance will not apply if the contractor had never before been licensed prior to the infraction. It only applies when a previously valid license has lapsed for bureaucratic reasons. An unlicensed California contractor is not barred from submitting a bid but it must acquire a valid license before commencing any work.

16.4 Entitlement

Our federal government is the single largest purchaser of design and construction services. Its enormous purchasing power is leveraged to implement federal policy goals. Among the most significant policy drivers impacting the construction industry are prevailing wages, set-asides for contractors who qualify as small businesses, affirmative action mandates, and domestic materials purchasing preferences.

Prevailing Wages

The **Davis-Bacon Act** applies to all contracts for federal construction projects (Davis-Bacon Act, 1931). The provisions of the Act require all contractors to pay their laborers and mechanics no less than the locally prevailing wages and fringe benefits for corresponding work on similar projects in the area. The Davis-Bacon Act directs the Department of Labor to determine such locally prevailing wage rates. Related legislation applies these prevailing wage and benefits provisions to any state or local public project for which federal agencies assist through grants, loans, loan guarantees, and insurance.

In its application, this has the effect of requiring all non-union contractors to pay prevailing union wages. For prime contracts in excess of $100,000, contractors and subcontractors must also, under the provisions of the Contract Work Hours and Safety Standards Act, pay laborers and mechanics, including guards and watchmen, at least one and one-half times their regular rate of pay for all hours worked over 40 in a workweek (Contract Work Hours and Safety Standards Act, 1962). The overtime provisions of the Fair Labor Standards Act may also apply to Davis-Bacon Act covered contracts (Fair Labor Standards Act of 1938 [FLSA], 1938).

Set-Aside

To implement the policy goal of fostering the growth of small businesses, Congress has set a goal for participation by small business concerns of not less than 23 percent of the total value of all prime contract awards for each fiscal year (Small Business Act, §644(g)(1)(A)(i), 1958). The FAR defines small business concern as,

> "...any business entity organized for profit (even if its ownership is in the hands of a nonprofit entity) with a place of business located in the United States or its outlying areas and that makes a significant contribution to the U.S. economy through payment of taxes and/or use of American products, material and/or labor, etc. [Small business concern] includes but is not limited to an individual, partnership, corporation, joint venture, association, or cooperative..." (FAR, §19.001, 1984).

The **U.S. Small Business Administration [SBA]** establishes size limitations for construction companies based upon annual revenues. Most construction companies cannot exceed average annual revenues in excess of $31 million to $32.5 million and still qualify as a small business concern. Specialty trade contractors are limited to $13 million in average annual revenues (Kelleher, 2008, p. 125). Small businesses concerns generally self-certify themselves. Disciplinary actions for misrepresentation include suspension, debarment from government contracting, fines up to $500,000, and up to ten years in prison for the most egregious violations.

Federal agencies are authorized to reserve a fair proportion of solicitations for small business concerns. These solicitations are known as **set-asides**. A federal agency may solicit set-asides directly or establish set-aside provisions within the bidding requirements of general contractors who are solicited to construct projects that are too large for the bonding limitations of small contractors. In the later case, these general contractors are required to provide subcontracting plans. These plans must state percentage goals for small subcontractors. Rejection of a subcontracting plan by the contracting officer will generally disqualify a bid as nonresponsive.

The general contractor to whom a project is awarded must exercise a good faith effort to recruit enough 1st tier subcontractors to meet their percentage goals. Suitable documentation must be submitted to justify any failure to meet those goals. Unjustified failure to meet small business entity goals is construed as a material breach of contract with termination for cause for which the general contractor acquires liability for damages.

Section 8(a) Program

The SBA can elect to contract directly with a federal agency and then subcontract that work to small business concerns. The authority for this is found in Section 8(a) of the Small Business Act and the program is named accordingly. To participate in a Section 8(a) program a small business entity must be both economically and socially disadvantaged enterprises. Economically disadvantaged enterprises are marked by diminished capital and credit opportunities compared to other contractors who are not economically disadvantaged.

Socially disadvantaged business enterprises [DBEs] include minority business enterprises [MBEs] (Black Americans, Hispanic Americans, Native Americans, and Asian Americans), women-owned business enterprises [WBEs], veteran-owned business enterprises [VBEs], and service-disabled veteran-owned [SDVO] business enterprises. To qualify as a DBE, a principal of the enterprise must demonstrate both an ownership stake of at least 51% and actual management control over the enterprise.

To be entitled under Section 8(a), the net worth of the qualifying owner must be less than $250,000, not including the value of his or her personal residence. Qualified enterprises must prove their eligibility every year and they may remain in the program for no more than nine years. To remain eligible the owner's equity cannot grow to exceed $750,000. The enterprise must also have a viable business plan and two years of prior experience in the type of work being awarded that demonstrates a reasonable probability of success in sustaining its business operations.

Small Disadvantaged Business Program

Another federal entitlement for small contractors is the Small Disadvantaged Business [SDB] program. Contractors who qualify for the Section 8(a) program will also qualify for the SDB program, except that the maximum net worth for entitlement is raised to $750,000 and there is no limitation to how many years it can remain in the program. SDBs are not required to demonstrate a reasonable probability of success and recertification occurs at three-year intervals.

SDB participants receive price evaluation adjustments for specific non-civilian agency solicitations where the industry sector has been marked by persistent and significant underutilization of minority firms and a demonstrated incapacity to alleviate the problem by other mechanisms (Kelleher, 2008, p. 136). Large contractors may include SDB participants in the proposals in order to avail themselves of price evaluation adjustments.

Other Notable Set-Asides

The **Historically Underutilized Business Zone [HUBZone]** program was established under the Small Business Reauthorization Act of 1997 (Small Business Reauthorization Act of 1997, 1997). The purpose of the HUBZone program is to facilitate investment in low-income areas, nonmetropolitan counties with high unemployment rates, Indian reservations, and other specifically designated localities.

Among other criteria, HUBZone contractors must show that 35% of their employees reside in a HUBZone area. HUBZone contractors must apply for certification - self-certification is not allowed – and recertify every three years. HUBZone contractors have priority over other types of set-asides and contracts may be awarded on a sole-source basis. Price evaluation adjustments are awarded in order to put HUBZone contractors on par with other SDB-qualified contractors.

The Veterans Benefit Act of 2003 established a set-aside program for small business **service-disabled veterans-owned [SDVO]** business enterprises. As with the HUBZone program, SDVO contractors must apply for certification; they have priority over other types of set-asides; contracts may be awarded on a sole-source basis; and price evaluation adjustments are awarded in order to put SDVO contractors on par with other SDB-qualified contractors.

Affirmative Action

The Office of Federal Contract Compliance Program (Office of Federal Contract Compliance Program, 1978) is charged with shielding individuals with protected classes characteristics from employment discrimination pursuant to federal construction projects. All federal contractors must pledge not to discriminate on the basis of race, color, religion, gender or national origin.

Federal solicitations may also stipulate an affirmative action plan establishing quotas for hiring in specified classes. For example, the Vietnam Era Veterans Readjustment Assistance Act of 1974 stipulates affirmative action goals for disabled veterans and veterans of the Vietnam era; and the Rehabilitation Act of 1973 stipulate affirmative action goals for individuals with disabilities (Vietnam Era Veterans Readjustment Assistance Act of 1974, 1974; Rehabilitation Act of 1973, 1973).

Executive Order No. 11141 prohibits age discrimination unless a bona fide occupation qualification can be demonstrated or statutory laws conflict with this order (FAR, §22.901(c), 1984).

Domestic Materials Purchasing

The **Buy America Act of 1988** stipulates the purchase of domestic materials for all federal construction projects (Buy America Act of 1988, 1988). Contracting officers are authorized to do otherwise if specific extenuating circumstances can be demonstrated, among which are unavailability of domestic supply, unreasonable prices, inadequate quality of supply, or other factors that can be demonstrated as against the public interest. The **Trade Agreements Act of 1979** enables the President of the United States to waive domestic procurement requirements pursuant to U.S. trade agreements on project that exceed $7.4 million in value (Trade Agreements Act of 1979, 1979).

Chapter 17 ENVIRONMENTAL LAW

17.1 Introduction

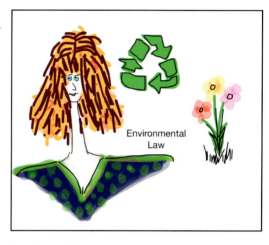

Federal environmental laws touch every facet of construction. Contractors not only must comply with federal environmental regulations, they are called upon to manage green and sustainable building programs that address energy use, water use, construction materials, waste reduction, and the indoor environment. Contractors are increasingly being called upon to facilitate brownfield programs aimed at abandoned, idled, or underused industrial and commercial facilities where expansion or redevelopment is complicated by real or perceived environmental contamination.

Contractors are responsible for ensuring that the many federal environmental regulations for construction projects are being met (Schierow, 2013). It doesn't matter whether the contractor is a subcontractor for clearing, grading or excavation; a demolition subcontractor; a plumbing, HVAC, electrical specialty contractor, a general contractor, a designer, or an engineer; everybody involved in the building process must exercise due diligence on environmental regulations. Due diligence is the exercise of a continual effort to accomplish something. In the common law, due diligence requires the care, caution and attention that would be reasonably expected of a person who is satisfying a legal obligation.

We are exposed to large numbers of chemical substances and mixtures each year. The United States Congress, in the **Toxic Substances Control Act [TSCA]**, asserts that: "Among the many chemical substances and mixtures which are constantly being developed and produced, there are some whose manufacture, processing, distribution in commerce, use, or disposal may present an unreasonable risk of injury to health or to the environment" (Toxic Substances Control Act [TSCA], §2601 (a)(1-2), 2014). The Toxic Substances Control Act [TSCA] established regulations for known toxic

CONSTRUCTION REGULATIONS

substances. Toxic substances that may be encountered in construction include asbestos, radon, and lead-based paints, and polycarbonated biphenyl [PCB].

As a part of the effort to control water pollution, the **Clean Water Act** regulates the discharge of storm water from any construction sites larger than one acre (Clean Water Act of 1977 [Clean Water Act], 1977). All construction projects on a one-acre or larger site must prepare a storm water pollution prevention plan [SWPPP].

17.2 Asbestos

Asbestos is actually a generic name for a group of six naturally occurring silicate minerals. Deposits of these minerals are mined and then processed into fibers for use in manufactured products. Widespread use of asbestos is attributable to its highly desirable physical properties such as heat resistance, chemical resistance, electrical resistance, sound absorption, tensile strength and durability. Its use in building materials is primarily for its flame retardant and insulating properties. Asbestos has been used in insulation, paint, plaster, acoustic ceilings, flooring, pipes, valves, fittings, gaskets, roofing shingles, siding, and many, many other building materials.

Asbestos is now known to cause asbestosis (scarring of lung tissues that results in constricted breathing), lung cancer, mesothelioma (cancer of the linings of the lungs and abdomen), and several other serious illnesses in humans. Exposure occurs when asbestos fibers, released into the air, are inhaled. These fibers are microscopically small and invisible to the naked eye. They are a significant health threat because human lungs do not have the ability to filter out such small particles. As a result, fibers can penetrate deep within the lungs where they can irritate and scar lung tissues. Even short-term exposure can be harmful and symptoms may not become apparent for years after exposure. Researchers believe that there is no safe level of exposure, although the amount of asbestos you are exposed to, the length of time you were exposed, and the number of times you were exposed, all impact your risk of illness.

Asbestos Risk

An **asbestos-containing material [ACM]** does not pose a health hazard when it is in good condition, intact and undisturbed. ACM becomes a health hazard once it is disturbed and its asbestos fibers are released into the air.

Some ACM is **friable**, meaning that it easily crumbles. These are the most hazardous forms because asbestos fibers easily become airborne when friable materials are impacted or when they deteriorate with age. Friable ACM was often sprayed or troweled onto ceilings and walls for thermal, acoustical, and decorative purposes. Other friable applications include insulation for stoves, furnaces, boilers, pipes, walls, and ceilings; and textured paints.

Some ACM contains asbestos fibers that are imbedded in place with a binder. These bound ACMs are somewhat safer. Some roofing shingles and siding incorporate asbestos fibers in a Portland cement binder. Because shingles and siding are installed outdoors they present little risk to human health. Vinyl-asbestos floor tiles, sheet flooring and artificial fireplace logs are all bound ACM and they are installed indoors where they present more risk. Nevertheless, all bound ACM can release asbestos fibers into the air when it is cut, drilled, scraped, or sanded.

The use of ACM in building products was phased out in the mid-1980s throughout most of the industrialized world. The **United States Environmental Protection Agency [EPA]** announced a phased-in ban of most asbestos products, but it was never implemented. The addition of asbestos to building materials has never been completely banned either in the United States or internationally. Regardless, costly class action litigation in the U.S.A. has had the practical effect of motivating insurance underwriters, financial institutions, and manufacturers to ban the use of ACM.

Current federal regulation of asbestos dates back to the 1984 **National Emission Standards for Hazardous Air Pollutants [NESHAP]** (1984). We know that ACM will likely still be found in structures built before 1985 and almost certainly will be present in those built before 1950. But ACM may be present even in recently built structures. Building products containing asbestos can still be found in the world and building materials recycled from scrapyards might contain just about anything. It is simply impossible to know what goes into every structure.

California Regulatory Scheme

In California, asbestos work must comply with relevant sections of Title 8 of the California Code of Regulations (California Occupational Safety and Health Regulations [Cal/OSHA], 2014). Work that disturbs ACM may also trigger compliance with other regulations of the EPA, the California Department of Transportation, the California Department of Health Services, the Contractor's State Licensing Board, and other local agencies.

CONSTRUCTION REGULATIONS

Today, the **California Environmental Protection Agency [Cal-EPA] Air Resources Board** enforces compliance with federal NESHAP regulations of asbestos and investigates all related complaints. The Air Resources Board delegates specific enforcement, however, to thirty-five geographically organized air districts within the state (California Environmental Protection Agency Air Resources Board, 2014).

Renovation Projects

Renovation is defined as any repair, maintenance or remodeling activity on any structure. Before any renovation is undertaken, all of the building materials that might be disturbed must be identified. If the contractor suspects that any of the identified materials contain asbestos a California-licensed asbestos consultant must be called in to conduct a survey. Licensed consultants must be Cal/OSHA certified and must have taken and passed a Cal-EPA approved building course. The contractor must also notify the local air district of the pending survey. Although an asbestos permit from the state is not required, some California cities require that contractors submit proof of notification to the air district before they will issue a renovation permit. The consultant is also required to file a survey report with the local air district.

The consultant may take samples of possible ACMs and send them to an accredited laboratory for analysis. The National Institute of Standards and Technology Asbestos Program accredits laboratories in California for analysis of asbestos in bulk building materials. If any test comes back positive for ACM, a decision must be made. If, in the structure to be renovated, the ACM is in good condition then the best course of action is to leave it alone. Bound ACM in particular, does not present a health hazard as long as it is not disturbed. But if the ACM is not in good condition, or it would be disturbed during the work, it must be completely removed from the structure.

To remove any amount of ACM in a safe and legal manner a California-licensed asbestos abatement contractor must be called in. Only asbestos removal contractors listed at the Cal/OSHA Asbestos Registration are allowed to remove ACM in the State of California. State inspectors may visit the work site before or during abatement to verify compliance with asbestos removal procedures and after the abatement is complete they must confirm that the ACM was removed prior to the start of renovations.

There are certain exceptions to the survey and notice requirements for homeowners working on their own home and for small ACM removal jobs.

Because of the many, complex, interdependent regulations, many contractors choose to avoid this work.

Demolition

Regardless of the date of a structure's construction, and because of the possibility that unreported renovations might have occurred during the life of the structure, an asbestos survey report is required prior to any demolition work. The survey report determines the presence or absence of ACM. A California-licensed asbestos consultant must be called in to conduct a survey and issue the survey report; the local air district must be notified; and the notification form must be posted on site. The California Health and Safety Code prohibit cities from issuing a demolition permit until they have been presented with a copy of the notification (California Health & Safety Code [Cal HSC], 1990). A California-licensed asbestos contractor must remove all ACM before the demolition starts. Only asbestos removal contractors listed at the Cal-OSHA Asbestos Registration are allowed to remove ACM in the State of California. State inspectors may visit the work site before or during abatement to verify compliance with asbestos removal procedures. And after the abatement is complete they must confirm that all of the ACM is removed prior to the start of demolition.

17.3 Radon

Radon is a colorless, odorless gas that seeps into buildings through cracks in walls and foundations. The gas is naturally produced as uranium decays in soil. Uranium is present in most rock and soil. It makes it way into the water table as it decays. According to the EPA, radon exposure kills more than 21,000 Americans each year (United States Environmental Protection Agency [EPA], n.d.(a)). To address that, the U.S. Congress passed the Indoor Radon Abatement Act in 1988, which allocated millions of dollars to, among other things, the development of model construction standards. Despite that, the United States may not yet have achieved the level of voluntary action required to make a real impact on indoor radon exposure.

California's Indoor Radon Program

California has long established an indoor radon program that seeks to reduce the health risk from radon exposures by increasing awareness, providing information about testing, identifying areas of high radon potential, and promoting mitigation and radon-resistant building techniques. Radon can be found throughout California. But although higher radon levels can be identified to some geographical areas more than others the amount of radon present in any house seems not to be related to the local geography.

Radon levels are detected by testing. Homeowners can buy inexpensive home test kits for less than $10. But contractors should subcontract to a radon tester certified by the State of California to conduct a radon test. Certification and registration of radon testers is a requirement in California. The California Department of Public Health along with the EPA recommend mitigating for radon if a building has an exposure of four (4) picocuries per liter (4 pCi/L) or higher.

Radon-Resistant Building Techniques

The EPA recommends that contractors of new homes use radon-resistant building techniques. Specific techniques vary for certain foundation types and difficult site conditions. However, there are five basic techniques - requiring no special skills or materials - that can be applied to the construction of most new homes (EPA, n.d.(a)).

- Gravel – Place a 4-inch layer of clean, coarse gravel below the slab in slab-on-grade construction. This prevents soil gases from building up by allowing them to escape from underneath the house.
- Vapor Retarder – Place 6 mil. or more of polyethylene sheeting below the slab in slab-on-grade construction or on top of the soil or gravel in a crawl space.
- Vent Pipe. – Run a 3-inch or 4-inch solid PVC Sch. 40 pipe continuously from the gravel layer (or from the crawl space) up through the roof. This works much like a plumbing soil stack to safely vent gases outside and above the house.
- Sealing and Caulking – Carefully seal all openings, cracks, and crevices in the slab-on-grade with polyurethane caulk to prevent radon and other soil gases from entering the home. Likewise, carefully tape all seams of the vapor retarder where there is a crawl space.
- Vent Fan – Install an electrical junction box or receptacle in the attic to use with a whole-house vent fan.

17.4   Lead-Based Paint

The term **lead-based paint [LBP]** means paint or other surface coatings that contain lead in excess of 1.0 mg/cm^2 or 0.5% by weight. House paints used to be made of three components: linseed oil, a solvent, and a pigment. The pigment is responsible for the color of paint. Lead oxide is a white pigment used in paint since ancient times. Lead chromate pigments in colors of yellow, orange or green (when mixed with a blue pigment) are also quite prevalent. Lead-based pigments were preferred over many other pigments because lead has greater hiding power, durability, moisture resistance, it speeds drying time, retains a fresh look longer and spreads easily.

Prior to 1940, almost all paints contained lead. As titanium dioxide, a white pigment of hiding power superior to lead oxide, became economical, the use of lead oxide diminished. By 1978 the use of lead oxide had all but stopped. Lead chromate paint is still used for safety paints, such as the paint on traffic lines or fire hydrants.

Lead may make good paint but the metal has no known biological benefit to humans. When lead is ingested into the body it creates health problems. Too much lead can damage various organs and systems including the nervous system, reproductive systems and the kidneys. It can cause high blood pressure and anemia. It interferes with the metabolism of calcium and Vitamin D. When lead builds up in the body - over a period of months or years - it accumulates in the bones and leads to lead poisoning. Lead is especially harmful to the developing brains of fetuses, young children and to pregnant women. High blood lead levels in children can cause consequences that may be irreversible including learning disabilities, behavioral problems, and mental retardation. At very high levels, lead can cause convulsions, coma and death.

Lead gets ingested into the body when LBPs are swallowed or inhaled. Ingestion can easily happen during the renovation of old, painted structures. When LBP is dry, sanding, scraping, burning, brushing or sandblasting disturbs it and workers may breathe in the lead dust or fumes. People in the vicinity of renovations, especially children, can unwittingly ingest lead dust or fumes as they eat, play and conduct other normal activities. Because of public health concerns, the U.S. Congress and the legislatures of the various states have caused regulations to be put in place that protect the public from the hazards of lead.

Federal RRP Program

Anybody who renovates structures built before 1978 needs to pay close attention to federal and state laws related to LBP. The EPA's **Lead-Based Paint Renovation, Repair and Painting Program [RRP]** of 2008 (as amended in 2010 and 2011) is a federal regulatory program aimed at contractors and others who disturb old, painted surfaces in the course of renovation work (EPA, 2011b). Renovation is broadly defined as any activity that disturbs painted surfaces and includes most repair, maintenance, and remodeling activities, including window replacement. It applies only to target housing. The term **target housing** is defined in the TSCA as any housing constructed prior to 1978, except housing for the elderly or persons with disabilities (unless any child who is less than 6 years of age resides or is expected to reside in such housing for the elderly or persons with disabilities) or any zero-bedroom dwelling.

The RRP aimed at protecting against lead-based paint hazards associated with renovations (EPA, 2011b). The RRP program develops and provides pre-renovation requirements, training, certification, and required work practices. Among other features of the RRP program: a lead information pamphlet must be distributed before starting renovation work; contractors are required to be certified; contractors' employees must be trained in use of lead-safe work practices; and lead-safe work practices that minimize occupants' exposure to lead hazards must be followed.

The RRP program must be complied with by anyone who is compensated to perform work that disturbs paint in target housing. These parties may include, but are not limited to residential rental property owners/managers; general contractors; and specialty subcontractors such as painters, plumbers, electricians, and carpenters. In general, any paid activity that disturbs paint in target housing is subject to the RRP program. Typical activities include remodeling, repair, maintenance, electrical work, plumbing, painting preparation, carpentry, and window replacement. Housing that has been declared lead-free by a certified lead abatement inspector or risk assessor is excluded from the program as is minor repair and maintenance activities. Minor repair and maintenance activities are those that disturb six square feet or less of paint per room inside, or 20 square feet or less on the exterior of a home or building. Minor repair and maintenance activities do not include window replacement and projects involving demolition or prohibited practices.

Pre-Renovation Education Requirements

In housing built before 1978, the contractor must distribute EPA's lead pamphlet to the owner and occupants before renovation starts. In a child-occupied facility, the contractor must distribute the lead pamphlet to the owner of the building or an adult representative of the child-occupied facility before the renovation starts; and either distribute renovation notices to parents or guardians of the children attending the child-occupied facility or post informational signs about the renovation or repair job. For work in

ENVIRONMENTAL LAW

common areas of multi-family housing, contractors must either distribute renovation notices to tenants or post informational signs about the renovation or repair job. Informational signs must: be posted where they will be seen; describe the nature, locations, and dates of the renovation; and be accompanied by the lead pamphlet or by information on how parents and guardians can get a free copy. Contractors must obtain confirmation of receipt of the lead pamphlet from the owner, adult representative, or occupants, or a certificate of mailing from the post office. Records must be retained for three years.

Pre-renovation education requirements such as the lead pamphlet, renovation notices, and information signs do not apply to emergency renovations. Emergency renovations are performed in response to a resident child with an elevated blood-lead level.

Training, Certification & Work Practice

All firms, including sole proprietors, performing compensated renovation work on target housing must be certified. All of the workers who are performing work for certified firms must be trained. Trained workers must follow lead-safe work practices. Examples of lead-safe work practices include: work area containment to prevent dust and debris from leaving the work area; prohibition of certain work practices like open-flame burning and the use of power tools without High-Efficiency Particulate Air [HEPA] exhaust control; and thorough clean up followed by a verification procedure to minimize exposure to LBP hazards.

California's Lead-Based Paint Rules

LBP rules in the State of California are derived from the California Code of Regulations (Accreditation, Certification and Work Practices for Lead-Based Paint and Lead Hazards, 2008). California LBP rules equal and exceed the federal LBP rules that are expressed in the RRP program. California LBP rules are more restrictive than the federal LBP rules in the following ways:

- EVERYONE must comply, not just those who are being compensated for their work;
- ALL structures built before 1978 are implicated, not just target housing;
- You must PRESUME that any untested paint or surface coating in a pre-1978 structure is LBP; and

CONSTRUCTION REGULATIONS

- There is NO *DE MINIMIS* for renovation activities. (e.g. RRP-specified activities that disturb six square feet or less of paint per room inside, or 20 square feet or less on the exterior).

In their combined effect, this says that any time a renovation is performed in a pre-1978 structure, no matter the type of structure, no matter how small the job, and regardless of whether there is compensation involved, you must presume that all surfaces are covered with LBP. California LBP work practice standards are also more stringent than the work practice standards found in the federal RRP program.

Highlights of the California LBP work practice standards are summarized, as follows:

- To have or to create a lead hazard is a violation of California Law. A lead hazard is defined as deteriorated LBP, lead contaminated dust, lead contaminated soil, disturbing LBP without containment, disturbing presumed LBP paint without containment, or any other nuisance attributable to lead. There is no *de minimus*. It is illegal to have or to create any lead hazard, no matter how small.
- A California-certified lead inspector/assessor must conduct a lead inspection and risk assessment. These must be documented in a lead hazard evaluation report. The original evaluation must be retained for a minimum of three years and copies distributed to the owner and to the California Department of Public Health.
- ONLY a California-certified lead inspector/assessor or a California-certified sampling technician may take samples. A California-certified sampling technician can only conduct visual inspections and take samples in the specific locations that are identified by a California-certified inspector/assessor. A certified RRP contractor cannot take samples. There is an exception for building materials that had already been removed from a structure and awaiting disposal, but only for testing to determine if hazardous waste requirements apply.
- Paint, dust, and soil samples taken for laboratory analysis must be analyzed by a laboratory that is recognized by the EPA (TSCA, §2685(b), 2002).
- ONLY a California-certified lead supervisor or a California-certified lead worker may perform abatement. A California-certified lead supervisor, California-certified lead project monitor, or California-certified lead project designer may be involved in planning and documentation. If any amount of known LBP or presumed LBP

is disturbed in a California structure, abatement personnel must: contain the work area; prepare an abatement plan; post notifications; employ lead-safe work practices; assure that there is no visible dust or debris at the end of the project; be prepared to demonstrate compliance with containment and lead-safe work practices; and maintain records for at least three years.
- The California-certified lead inspector/assessor, California-certified lead project monitor, and California-certified lead sampling technician conducting the lead hazard evaluation cannot conduct abatement on the same structure.
- After abatement work is completed a California-certified lead inspector/assessor or a California-certified lead project monitor must conduct a clearance inspection.

17.5 Polychlorinated Biphenyls

Polychlorinated biphenyls [PCBs] are any chemical substance with a biphenyl molecule that has been chlorinated to varying degrees or any combination of substances that contain such a substance. Due to their non-flammability, chemical stability, high boiling point, and electrical insulating properties, PCBs were used in hundreds of industrial and commercial applications including electrical, heat transfer, and hydraulic equipment; as plasticizers in paints, plastics, and rubber products; in pigments, dyes, and carbonless copy paper; and many other applications.

PCBs have been demonstrated to cause a variety of adverse health effects. PCBs have been shown to cause cancer in animals. PCBs have also been shown to cause a number of serious non-cancer health effects in animals, including effects on the immune system, reproductive system, nervous system, endocrine system and other health effects. Studies in humans provide supportive evidence for potential carcinogenic and non-carcinogenic effects of PCBs. Items with PCB concentrations of 50 parts per million (ppm) or greater are regulated for disposal under 40 CFR Part 761. The most likely sources of PCBs at construction sites are:

- Capacitors or transformers manufactured prior to July 2, 1979;
- Fluorescent light ballasts manufactured prior to July 2, 1979;
- Mineral-oil filled electrical equipment such as motors or pumps manufactured prior to July 2, 1979;
- Waste or debris from the demolition of buildings and equipment manufactured, serviced, or coated with PCBs;

CONSTRUCTION REGULATIONS

- Plastics, molded rubber parts, applied dried paints, coatings or sealants, caulking, adhesives, paper, sound-deadening materials, insulation, or felt or fabric products such as gaskets manufactured prior to July 2, 1979; and
- Waste containing PCBs from spills, such as floors or walls contaminated by a leaking transformer.

For demolition activities, the building owner should have an inventory of all items containing PCBs. However, if the equipment is not marked and if no records exist, any PCB-containing materials need to be identified. PCBs are difficult to locate and identify. PCB concentrations may vary from item to item and within classes of items. Specialists in environmental engineering, remediation, or sampling and analytical services are called upon to identify PCBs.

Federal PCB regulations define the <u>generator</u> as being responsible for handling, storing, transporting, and disposing of PCB wastes (PCBs Manufacturing, Processing, Distribution in Commerce, and Use Prohibitions, 2014) The generator is considered the party that owns the material. For most construction projects, multiple parties will be involved; all may be liable if the PCB handling and disposal requirements are not followed.

In a typical construction project, PCB wastes are generated in one of two ways:

- PCB-contaminated soils and materials are discovered during grading or digging (i.e., remediation wastes); or
- PCB-contaminated buildings or equipment are discovered during demolition.

The contractor or subcontractor who first discovers the PCB-containing material typically is responsible for notifying the general contractor, developer, and/or owner. Because the PCB-containing material was present on the site prior to construction activities, the developer or owner typically is responsible for ensuring that all PCB wastes are handled and disposed of properly.

Because PCBs were banned as of 1979, it is unlikely that a contractor or subcontractor will bring PCB-bearing materials on site during a construction project. However, if this happens, that contractor or subcontractor typically would be responsible for complying with the PCB requirements.

17.6 Storm Water Pollution Prevention Plans

As a part of the effort to control water pollution, the **Clean Water Act** regulates the discharge of storm water from any construction sites later than one acre (Clean Water Act of 1977 [Clean Water Act], 1977). All construction projects on a one-acre or larger site must prepare a storm water pollution prevention plan [SWPPP].

A SWPPP plan must be submitted to the EPA along with a notice of intent, prior to commencing any construction work on site. The EPA will issue a permit if the plan meets EPA requirements including descriptions of storm water control mechanisms; methods for maintaining those control measure; and the identities of the contractors and subcontractors that will implement the plan (Kelleher, §17.9, 2008). At completion of construction work a notice of termination of the permit is filed, asserting that storm water discharges have either ended or that the contractor is no longer responsible for the site.

Violations of the Clean Water Act may receive civil penalties of up to $25,000 per day, criminal penalties of up to $25,000 per day, and up to one year in prison.

Chapter 18   LABOR LAW

18.1   Introduction

The industrial revolution brought about the rise of centralized manufacturing, with factories replacing the cottage industry in which craftsmen produced their own goods. The laborers who worked in these factories were subjected to harsh conditions and long hours. The laborers finally realized they had to join to-

Labor Law

gether so that not just the employers would benefit from the improved standard of living. The workers wanted to ensure that they would get their share of the increasing wealth of the nation. The employers originally went to the courts (and ruling class) to create laws so that the laborers were held as illegal (in their labor groups) or conspirators. Joint action by workers eventually won out and was recognized as legitimate.

18.2   The U.S. Labor Movement

The courts and the law were major obstacles to achieving a legitimate place in society for organized labor. The first obstacle was having the law recognize that a labor organization was not per se an illegal conspiracy. That legal development came as a decision of the Massachusetts Supreme Court in 1842 which held that union activities were not illegal conspiracies as long as the objectives of those activities, and the means employed to achieve those objectives, were not illegal. (*Commonwealth v. Hunt*, 1842). Afterward, the legality of labor unions was accepted by mainstream judicial opinion.

Still the labor movement struggled with employers for recognition. The last decades of the nineteenth century were marked by violent strikes in several industries. There were three centers of labor that attempted to rejuvenate organized labor during this time period and two centers of labor emerged from those during the mid-to-late twentieth century. These labor groups included:

The Knights of Labor. First developed in Philadelphia in 1869, this organization originally sought to stand up for both skilled and unskilled workers. After the Railway Strike of 1877 the Knights became a national organization and between 1878 and 1884 they conducted a large number of strikes. Because their focus was on industry-wide organization rather than craft unions, the unskilled workers suffered at the hands of their employers who could easily replace the unskilled workers during a strike. At times, following unsuccessful strikes, local workers were forced to sign **yellow-dog contracts** in which they agreed not to join any union. Eventually the skilled craft unions within the Knights of Labor came to believe that they could more effectively achieve their goals through narrowly based organizations emphasizing labor actions rather than political efforts. Because of this there was a decline in membership of the Knights when the craft unions pulled out.

The Socialists. The German Workingmen's Union (later to be reorganized as Section 2 of the First International which was commonly known as the Social Party) was formed in New York City in 1865, based on the establishment in 1864 of the International Workingmen's Association (the First International) by Karl Marx in London. After the Railway Strike of 1877 it turned to political activities with its political arm becoming the Socialist Labor Party. The public stopped accepting the legitimacy of the socialist movement following the Haymarket Riot of 1886 in Chicago.

The labor activities of the socialist movement came to be represented by the Industrial Workers of the World [the IWW, or "Wobblies"] during the first decades of the twentieth century. The Wobblies were known for their violent strikes. The American Communist Party eclipsed the Wobblies following the Russian Revolution of 1917 and by the late 1940s and early 1950s the Cold War and McCarthy "red hunts" brought an end to organized labor's links to the American Communist Party.

The American Federation of Labor [AFL]. This labor group started by Samuel Gompers and Adolph Strasser of the Cigarmakers' Union, ultimately became the dominant organization of the American labor movement because it emphasized union activities in contrast to political activities. Based on the British trade union system, the AFL adopted the following "ideas":
- Local unions were to be organized under the authority of a national association (whose focus was on wages and practical, immediate goals).

- Dues were to be raised to create a large financial reserve.
- Sick and death benefits were to be provided to members.

By 1900 there were approximately 500,000 skilled workers in these AFL unions. The AFL and its affiliated craft unions dominated the organized labor movement in America with growing numbers of members (workers) during the first few decades of the twentieth century.

The Congress of Industrial Organizations [CIO]. The AFL had largely ignored the unskilled production workers (autoworkers, steelworkers, and mineworkers). John L. Lewis led these workers and created a federation of unions calling it the CIO. The AFL opposed this new organization, and in 1938 they expelled all unions associated with the CIO. The workers of the steel, automobile, rubber, electrical, manufacturing, and machinery industries had spectacular success working on political activities.

The AFL-CIO. After years of rivalry, the AFL was finally forced to recognize the CIO. In 1955, the AFL (by this time 10.5 million members) merged with the CIO (with 4.5 million members). The **AFL-CIO** was the dominant organization in the American labor movement.

Recent Trends

Union membership in the private sector reached a peak in the early 1950s with approximately one-third of the American labor force unionized. Since then, there has been a slow decline. By 2009, only about 12 percent of the work force (public and private sector) were unionized. The manufacturing sector of the U.S. economy has been hit hardest by the changing economic conditions and global competition.

American industry was restructured during the late 1970s and 1980s by mergers, takeovers, plant relocations (to the mostly non-union Sun Belt and overseas), plant closings, and collective bargaining where employers asked for, and received, reductions in wages and benefits and relaxation of restrictive union work rules. During the mid-1980s this created a decline in the U.S. manufacturing sector and the rise of the service economy.

This decline continued into the 1990s and the first decade of the 21st century. The ultimate outcome has been the decline of the U.S. middle class coupled with the outsourcing of jobs to low-wage countries. However, while private sector unions have been in decline, the public sector unions have been growing since the 1960s. By 2009 about 36 percent of government employees were union members.

Legal Responses

By the 1880s, employers wanted new legal weapons to use against labor activists. The U.S. legal system reacted to organized labor by initially trying to suppress it through the use of the labor injunction, the yellow-dog contracts, and the antitrust laws against union strikes and boycotts.

Injunctions. A powerful weapon developed in the late 1880s against the activities of organized labor; the injunction was utilized throughout the last decade of the nineteenth century and the first two decades of the twentieth century. An injunction is a court order directing a person to do, or to refrain from doing, specific actions. Once an injunction had been granted, court officers would enforce it against the union and union leaders generally had to comply by stopping the strike or boycott. Union members who resisted risked jail terms and/or fines for being in contempt of the court order. It became an effective weapon for management to use against any union pressure maneuvers.

Yellow-Dog Contracts. Yellow-dog contracts were employment agreements that would prohibit a worker from joining a union and that then became a condition of employment. If the union encouraged the workers to break their contracts, then the union was guilty of wrongly interfering with contractual relations. These yellow-dog contracts were an extremely effective weapon against unions.

Antitrust Laws. The **Sherman Antitrust Act** was the first of the antitrust laws. It outlawed restraints of trade and monopolizing of trade. Passed in 1890 by Congress, the act was designed to protect free trade and competition. However, employers used this act and other antitrust laws to challenge union strikes and boycotts by seeking damages for interference with trade.

18.3 The National Labor Relations Act

Historical Background

The Norris-LaGuardia Act. In 1932 a federal anti-injunction act sponsored by Senator Norris and Congressman LaGuardia was enacted (Norris-La Guardia Act, 1932). Several states passed similar legislation restricting the ability of their court systems to issue labor injunctions (known as "little Norris-LaGuardia Acts"). The act prohibited the federal courts from issuing injunctions in labor disputes except in strict compliance with the provisions set out in the act. Only after complying with the provisions can a court issue an injunction in a labor dispute. The act lists activities that are safeguarded from injunctions and declares that yellow-dog contracts are contrary to public policy of the U.S.; therefore unenforceable by any federal court. Nor can the courts use such contracts as the basis for permitting any legal or reasonable remedies (injunctions). Finally, the act defined the term **labor dispute**, which states:

> "The term "labor dispute" includes any controversy concerning terms or conditions of employment, or concerning the association or representation of persons in negotiating, fixing, maintaining, changing, or seeking to arrange terms or conditions of employment, regardless of whether or not the disputants stand in the proximate relation of employer and employee" (Norris-La Guardia Act, §113(c), 1932).

The National Industrial Recovery Act [NIRA]. This was the centerpiece of President Franklin D. Roosevelt's New Deal when he took office in 1933. The NIRA set up a system in which major industries would operate under codes of fair competition. Trade associations for each industry would develop codes. These associations would be under the supervision and guidance of the National Recovery Administration [NRA] The NRA announced that codes containing provisions concerning hours, rates of pay, and other conditions of employment would be subject to NRA approval. The practical effect of this announcement was to allow industry to develop such codes unilaterally, without input from organized labor. While employees rushed to join unions, employers refused to recognize and bargain with the unions. The result was an upsurge of strikes (Castagnera & Cihon, 2011, pp. 346-347).

The National Labor Board [NLB]. President Roosevelt created a seven member National Labor Board in August of 1933 to "consider, adjust and settle differences and controversies that may arise through differing interpretations" of the NIRA provisions. The original mission of his "mediation board" was to persuade the parties to settle their differences peacefully. This power of persuasion was effective only as long as an employer was not blatantly hostile to organized labor (Castagnera & Cihon, 2011, p. 347).

CONSTRUCTION REGULATIONS

The NLB made several contributions to modern labor law that included the following principles:

- Rule of majority;
- Representation of the employees in a particular bargaining unit;
- Obligation of employees to bargain with a union that had been chosen as representative by a majority of the employees;
- Denial of employers' right to know of an employee's membership in, or vote for, a union when a secret ballot representation election was held; and
- Security of employment and entitlement to displace replacement workers if the strike was the result of NIRA violations by the employer.

The NLB's effectiveness was destroyed, however, when President Roosevelt made a decision to have the NRA negotiate a settlement in a case involving automobile manufacturers and the United Automobile Workers Union (Castagnera & Cihon, 2011, p. 347).

The "Old" National Labor Relations Board. In June 1934 President Roosevelt ended the NLB and transferred its funds, personnel, and pending cases to the National Labor Relations Board (or the "old" NLRB). This board published guidelines to assist regional offices in handling common types of cases and began organizing its decisions into a body of standards guiding future action. Unfortunately, the "old" NLRB was denied all jurisdictions over clashes in the steel and auto industries. The end came in 1935 when the Supreme Court declared the NIRA to be unconstitutional. That also destroyed the "old" NLRB (Castagnera & Cihon, 2011, p. 348).

Executive Orders 10988 and 11491. The Wagner Act protects the collective bargaining rights of private sector employees but not of public sector employees. Executive Order 10988 issued by President John F. Kennedy in 1962, and later replaced by executive order 11491 issued by President Richard M. Nixon in 1969, corrected that by recognizing the collective bargaining rights of federal employees. These executive orders established the Federal Labor Relations Council and vested it with responsibilities similar to those of the NLRB.

Overview

The passage of the **National Labor Relations Act [NLRA]** (sometimes called the Wagner Act), constituted a revolutionary change in national labor policy. The NLRA empowered the federal government to protect the rights of private sector workers to organize for mutual aid and security and to bargain collectively through representatives of their own choice. New York Senator Robert F. Wagner introduced the NLRA in the Senate. It was passed by Congress and enacted into law in 1935. The basic purpose of the act is to protect the basic rights of private sector employees. To protect their basic rights the act forbids certain employer practices, known as **unfair labor practices**, that hinder or prevent the exercise of such rights (National Labor Relations Act [NLRA], 1935).

The five practices listed in the NLRA include:

- Interference with, or restraint, or coercion of employees in the exercise of their rights;
- Domination of, or interference with a labor organization (including financial or other contributions to it);
- Discrimination in terms or conditions of employment for the purpose of encouraging or discouraging union membership;
- Discrimination against an employee for filing a charge or testifying in a proceeding under the act; and
- Refusal to bargain collectively with the employees' legal bargaining representative (NLRA, 1935).

The NLRA preempts state laws that purport to regulate conduct protected by or prohibited by the NLRA and that seek to regulate areas left by the NLRA to the market and economic forces. Most notably, the NLRA has been amended several times by the following two acts.

The Taft-Hartley Act of 1947 – Formally titled the Labor Management Relations Act, the purpose and effect of this act were to balance the rights and duties of both unions and employers. It added a list of unfair labor practices by unions. It outlawed the closed shop, a term describing an employer who agrees to hire only employees who are already union members. Finally, it emphasized that employees had the right to refrain from collective activity as well as engage in it (Labor Management Relations Act, 1947).

The Landrum-Griffin Act of 1959 – Formally titled the Labor-Management Reporting and Disclosure Act of 1959, this act was passed in response to concerns about union racketeering and abusive practices aimed at union members. The act set out explicit rights for individual union members against the union, and it prohibited certain kinds of conduct by union officials (such as financial abuse, racketeering, and manipulation of union-election procedures) (Labor-Management Reporting and Disclosure Act, 1959).

The National Labor Relations Board

A **National Labor Relations Board** [NLRB or the "Board"] was re-formed to enforce and administer the NLRA. The original charge of the Board was to create a nationwide organization, develop a body of legal precedents, and develop and refine its procedures. Now, the Board settles complaints of unfair labor practices under the NLRA and conducts representation elections.

Organization

Initially, the Board combined both enforcement and judicial functions within its charge. Widespread criticism was leveled at the Board for having combined what were seen as conflicting functions. Passage of the Taft-Hartley Act addressed this problem. The Taft-Hartley Act retained the Board as the enforcement agency but gave the Office of the General Counsel - an independent agency - administrative authority over the Board's regional enforcement efforts. The Board retained its judicial functions. At the end of the day, the organization bifurcated into two independent authorities within a single agency, the Board and the General Counsel.

The Board is the judicial branch of the agency. Originally composed of three members, it was expanded to five with the Taft-Hartley Act. All five members of the Board are nominated by the U.S. president and must be confirmed by the Senate. The president also designates one of the five Board members to be its chairperson. Board members serve five-year terms. Members have a staff of about twenty-five legal clerks and assistants. The Board has an executive secretary (chief administrative officer) charged with ruling on procedural questions, assigning cases to members, setting priorities in case handling, and conferring with parties. Additionally, it has a solicitor (to advise members on questions of law and policy) and an information director (to assist the Board on public relations issues). Members of the Board can be removed from office by the president only for neglect of duty or malfeasance in office.

The Board also has a branch called the Division of Judges that oversees **administrative law judge [ALJ]** functions. Each ALJ is independent of both the Board and the General counsel. Appointed for life, they are subject to the federal Civil Service Commission rules governing appointment and tenure. They were formerly called trial examiners because they conduct hearings and issue initial decisions on unfair labor practice complaints issued by regional offices throughout the U.S.A. An ALJ functions as a specialized trial court judge to decide unfair labor practice complaints. An ALJ is not charged with enforcement.

The **NLRB Office of the General Counsel** is the prosecutorial branch of the NLRB. The president nominates the general counsel subject to Senate confirmation for a four-year term. The general counsel is also in charge of the day-to-day administration of the NLRB regional offices. The Office of the General Counsel has four divisions: operations management, advice, enforcement litigation, and administration. There are thirty-four regional offices and a number of sub-regional offices. The staff of each regional office consists of a regional director, regional attorney, field examiners, and field attorneys.

Jurisdiction

The NLRB can regulate labor disputes in virtually any company, unless the firm's contact with interstate commerce is *de minimus* (miniscule and merely incidental). It has formed standards for determining whether or not it has jurisdiction over a particular labor dispute. Jurisdictional standards are set in terms of the dollar volume of business that a firm does annually. The NLRB has jurisdiction over virtually every contractor, including multi-employer bargaining associations, and multi-state firms. The NLRB also has jurisdiction over craft unions whenever they operate as employers.

Certain employers and employees are exempted from NLRB regulation. Exempted employers include: the federal government or any wholly owned government corporation; any state or political subdivision thereof (county, local, or municipal governments); any companies that are subject to the Railway Labor Act; and labor organizations in their representational capacity. Exempted employees include: individuals employed as agricultural laborers; individuals employed as domestics within a person's home; individuals employed by a parent or spouse; independent contractors (a person working as a separate business entity); individuals employed by employers subject to the Railway Labor Act; and supervisors.

A supervisor is defined as someone who has the authority to direct, hire, fire, discipline, transfer, assign, reward, responsibly direct, suspend, or adjust the grievances of other employees and who uses independent judgment in the exercise of such authority in the interests of the employers.

Exemptions have also arisen from federal appellate court decisions. Judicially exempted employees include managerial employees and confidential employees.

18.4 The Unionization Process

The NLRA gives employees the right to determine for themselves whether they wish to be represented by a union. The NLRA asserts that, "employees are entitled to bargain collectively through representatives of their own choosing" (NLRA, §157, 1935). The position of the union as exclusive bargaining agent supersedes any individual contracts of employment made between the employer and the unit employees. The NLRA further asserts:

> "Representatives designated or selected for the purposes of collective bargaining by the majority of the employees in a unit appropriate for such purposes, shall be the exclusive representatives of all the employees in such unit for the purposes of collective bargaining in respect to rates of pay, wage, hours of employment, or other conditions of employment" (NLRA, §159(a), 1935).

The **bargaining unit** is central to the notion of labor relations under the NLRA. A bargaining unit is a group of employees being represented by a union. In order for collective bargaining to produce results that are just to both sides, it is essential to define the bargaining unit. A bargaining unit can be defined as a single employer, multiemployer unit, or a craft unit. Professional employees can only be included in a unit with nonprofessionals if a majority of the professionals vote for inclusion. Any dealings with the individual unit employees must be in accordance with the collective bargaining agreement established with a bargaining unit. The Taft-Hartley Act also added some protection for minority factions within bargaining units.

Union Certification

The NLRA requires that a union, in order to become the exclusive bargaining agent, be designated or selected by a majority of affected employees. It does not require that an election by secret ballot be held to determine employee choice. Employee choice can be established by **voluntary recognition**. As long as majority support is evident and uncontested, an employer may agree to recognize a union as the exclusive bargaining agent for its workers, without holding a representation election.

But, a **representation election** has several advantages over voluntary recognition which include:

- Employees can be grouped together into bargaining groups in the election process;
- Employers are obligated to recognize and bargain with the victorious union for at least twelve months following its election;
- Petitions seeking a new representation election cannot be filed during the subsequent twelve-month period; and,
- Employer and employees, in most cases, prefer a representation election conducted by the NLRB.

The NLRB will not proceed with an election until the petitioning union presents evidence that at least 30 percent of the employee group support the election request. The employer must, on request, give the union an **Excelsior list,** a tabulation of the names and addresses of the employees eligible to vote in a representation election. A showing of support is reflected in signed and dated **authorization cards** obtained by the union from individual, eligible employees. These cards indicate that the employees authorize the union to act as the employee's bargaining agent and to seek an election on behalf of the employees. Other acceptable showings of employee interest can include a letter (or other document) displaying a list of signatures and applications for union membership.

The NLRB requires that a union recording a petition for a representation election must supply evidence to support their petition within 48 hours of the filing. This is known as the **forty-eight hour rule**. The parties may agree to waive their rights to a hearing on these issues and proceed to a **consent election**. This is an election conducted by the regional office giving the regional director final authority over any disputes. If the parties fail to agree on some of these issues and have not agreed to a consent election, then a representation hearing will be held. After the hearing, the hearing officer submits a report to the director. The director then renders a decision either to hold an election or to dismiss the petition. The election usually occurs twenty-five to thirty days after it has been ordered.

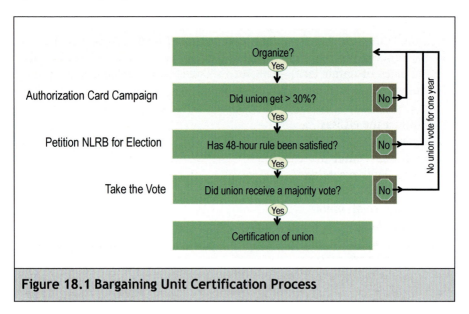

Figure 18.1 Bargaining Unit Certification Process

Once an election is ordered, both sides must maintain **laboratory conditions** to ensure that employees can exercise free choice. Neither the employer nor the union can do anything that would unduly influence employees' free choice. Violations of laboratory conditions include incentives, threats, and extreme third-party propaganda, other unfair labor practices, and captive-audience speeches. **Captive-audience speeches** consist of meetings/speeches held by the employer during working hours which employees are required to attend. Both sides must observe the **twenty-four hour silent period** immediately prior to the election. The union or employee group must refrain from formal campaign meetings during the twenty-four hour silent period.

Voting is accomplished by secret ballot on the day of the election. If the union does not receive a simple majority of the votes, the union does not become the certified bargaining agent for the employee bargaining unit and the NLRB will not accept any new representation election petitions for the next twelve months. But as long as a simple majority of eligible employees vote affirmatively, the union becomes the certified bargaining agent for all of the employees in that bargaining unit.

Upon certification, both the employer and the certified union are under a duty to bargain in good faith with each other, for a reasonable length of time, until a collective bargaining agreement is reached. No rival union may file a petition for a representation election during this time, up to a maximum of twelve months.

Unions may also obtain representation rights by means of unfair labor practice proceedings. The NLRB may issue a bargaining order when the consequences of unfair labor practices by employers prevent a fair election from being held. Such remedies are the exception. The NLRB and the courts prefer elections as the means to give effect to employees' right of free choice under the NLRA.

Once a labor agreement is reached, the contract-bar rule prohibits any new elections within the bargaining unit during the lifespan of the labor agreement. Two exceptions apply to this rule: 1) during the period from 60 to 90 days before the expiration of the labor agreement, known as the **open season**, any rival union can file a representation election petition; and 2) when the labor agreement is for more than three years, the contract-bar rule expires, allowing any rival union to file a representation election petition at any time after the end of the agreement's third year.

When the open season expires without a petition being filed by a rival union, the parties are free, within the current contract's final 60 days, to negotiate a new contract. But if they fail to reach agreement and their current contract expires, any rival union becomes free to petition for a representation election.

In the event that a union loses the support of a majority of the employees in a bargaining unit, the employees may file a **decertification petition** with the NLRB. As with the certification petition, this requires a showing of 30% of eligible employees in support of decertification followed by a simple majority vote if this is achieved.

Unfair Labor Practices

The NLRB handles two types of legal questions: representation questions and unfair labor practice complaints. Representation questions are concerned with whether, and if so, how employees will be represented for collective bargaining. These are known as "R" cases. Allegations that an unfair labor practice has taken place are known as "C" cases. The NLRB does not initiate proceedings under either question: rather, it responds to a petition for an election filed by a party to the case or a complaint of unfair practice.

CONSTRUCTION REGULATIONS

It is an unfair labor practice for an employer to:

1. Interfere with, restrain, or coerce employees in the exercise of their rights;
2. Dominate or interfere with the formation or administration of any labor organization or contribute financial or other support to it;
3. Discriminate in regard to hiring or tenure of employment or any term or condition of employment to encourage or discourage membership in any labor organization;
4. Discharge or otherwise discriminate against an employee because he has filed charges or given testimony with respect to labor practices; or
5. Refuse to bargain collectively with the representatives of his employees.

It is an unfair labor practice for a labor organization or its agents to:

1. Restrain or coerce
   A. Employees in the exercise of their rights; or
   B. Employer in the selection of its representatives for the purposes of collective bargaining or the adjustment of grievances;
2. Cause or attempt to cause an employer to discriminate against an employee or to discriminate against an employee with respect to whom membership in such organization has been denied or terminated on some ground other than his failure to tender the periodic dues and the initiation fees uniformly required as a condition of acquiring or retaining membership; or
3. Refuse to bargain collectively with an employer, provided it is the representative of his employees.

The consequence of unfair labor practices may be the deferral or discontinuance of an election, but a union may choose to proceed despite unfair labor practice charges. The NLRB may issue a bargaining order rather than proceed with an election. This could happen if the unfair labor practices were pervasive and outrageous, or if the union had majority support at some point.

18.5 Labor Disputes

When collective bargaining fails to produce an agreement in a labor dispute and the negotiations reach an impasse, either party may resort to coercive tactics to try to force the other side to settle the dispute. Coercive tactics might also inflict economic harm upon the other party.

The Norris-LaGuardia Act prohibits the federal courts from issuing injunctions in labor disputes against union conduct (Norris-LaGuardia Act, §107, 1932). The Act defines labor dispute broadly and includes strikes that are politically motivated. State courts can regulate violent picketing or picketing in violation of state laws.

18.5.1 Strikes and Lockouts

A **strike**, sometimes called a **walkout**, is usually an orderly, planned refusal of employees to work at a particular site, sometimes authorized by a vote among union members. This is the traditional weapon by which workers attempt to pressure their employers. A strike may happen over wages and compensation or because of unfair labor practices. Workers will sometimes walkout in a strike that is not authorized by their union or in opposition to the orders of their union leaders. This is known as a **wildcat strike**. There is no constitutional right to strike, but a strike is protected activity under the NLRA.

The NLRB distinguishes whether employees are on an **unfair labor practice strike** or an **economic strike**. An unfair labor practice strike is a strike to protest employer unfair practices. An economic strike is a strike over economic issues such as a new contract or a grievance.

Strikes by private sector employees are regulated by the NLRA and are protected activity under the Act (NLRA, §157, 1935). For public sector employees, there may be no lawful right to strike. **A lockout** is a refusal by an employer to allow workers to work at the jobsite. Employers initiate lockouts with the intent that withholding work from employees will force the union to make concessions. An employer may resort to this only after good faith bargaining has reached a stalemate.

It is common for labor agreements to have **no-strike, no-lockout provisions**. These provisions bar union workers from going out on strike and prohibit employers from refusing to permit them from working on site during a labor dispute. These provisions benefit contractors by assuring that their projects will not be delayed and unions invariably gain concessions in exchange for agreeing to these provisions.

CONSTRUCTION REGULATIONS

When, during a labor dispute, workers do not want to strike or are barred from striking by a no-strike clause, they may choose to slowdown. A **slowdown** is an intentional and methodical reduction in labor productivity at a job site. It is done to exert pressure on an employer during labor negotiations. A slowdown may not violate the letter of a no-strike provision but it clearly is a bad faith violation of its intent. Severe slowdowns may rise to the level of unfair labor practices. In any event, most observers would condemn slowdowns as unethical negotiating ploys.

18.5.2 Picketing

"Peaceful and truthful discussion" designed to convince others not to engage in behavior regarded as detrimental to one's own interest, or to the public interest, is fully protected speech under the First Amendment of the Constitution. Picketing is the act of placing persons outside an employer's premises to convey information verbally, via words, signs, or by distributing literature. It is employed as a tactic both during union organizing efforts and in labor disputes.

Peaceful picketing is protected activity under the NLRA and the NLRB's role in regulating picketing is limited. State courts may issue injunctions against picketing in certain situations. These include **violent picketing**, endangering the safety of citizens and in violation of criminal laws, **mass picketing**, in which pickets march so closely together that they block access to a plant or project site, and use of picket signs involving language that constitutes fraud, misrepresentation, libel, or inciting a breach of the peace.

Recognitional picketing is picketing by an uncertified union to develop support for it to become the exclusive bargaining agent for particular workers. The NLRA prohibits some types of recognitional picketing. It is prohibited if: (a) its purpose is to force an employer to recognize a union as representative of its employees when another union has already been legally recognized; (b) a representation election had been conducted within the preceding twelve months; or (c) the picketing occurs without the uncertified union having petitioned the NLRB for a representation election (NLRA, §158(b)(7), 1935).

Labor dispute picketing is picketing by a certified union to develop support for its goals in a labor dispute. It is usually set up in conjunction with a strike at a location where work is being performed. This location is referred to as the **situs** of the labor dispute. A construction site becomes a situs when it is the location where a contractor performs work that is the subject of a labor dispute. Labor dispute picketing is a lawful activity except when the picketing is construed as an illegal secondary boycott.

Secondary Boycotts

One of a labor union's objectives during a strike might be to dissuade other people and businesses not involved in the labor dispute from doing business with the contractor whom they are on strike against. Labor law refers to the struck contractor as the **primary contractor**. If others were to actually stop doing business with the primary contractor they would be described as conducting a primary boycott. A primary boycott could be encouraged through publicity, such as radio, television, or internet advertising; handbills; or public speeches that truthfully inform the public about the objectionable business practices of the primary contractor. Truthful publicity that encourages a boycott against a primary contractor is expressly protected by the NLRA and supported by the 1st amendment constitutional right of free speech (NLRA, §158(b)(4), 1935).

Another one of a labor union's objectives during a strike might be to dissuade other people and businesses not involved in the labor dispute from doing business with those businesses whom are doing business with the contractor that they are on strike against. Labor law would describe this as a secondary boycott. Secondary boycotts can be encouraged through acts of publicity or picketing or both but they might be prohibited by the NLRA. (NLRA, §158(b)(4)(B), 1935).

An unlawful secondary boycott has two elements. First, a labor organization must "threaten, coerce, or restrain" a person engaged in commerce (NLRA, §158(b)(4)(ii), 1935). Second, the labor organization must do so with "an object" of:

> "...forcing or requiring any person to cease using, selling, handling, transporting, or otherwise dealing in the products of any other producer, processor, or manufacturer, or to cease doing business with any other person." (NLRA, §158(b)(4)(ii)(B), 1935).

Both elements must be present to constitute a violation of the Act

Yet another one of a labor union's objectives during a strike might be to dissuade nonunion tradespeople (sometimes referred to with the derogatory expression, "scabs") from working for the struck contractor. This is can also be encouraged through acts of publicity or picketing or both. A refusal to work for the primary employer is a primary boycott. Union activities to induce this sort of primary boycott are not prohibited by the NLRA. But a primary contractor might work around a primary boycott by subcontracting to another contractor. Although the subcontractor in these circumstances is a secondary employer the courts have reasoned that because they are doing the work of the primary employer, a picket aimed at them is a lawful, primary picket. This ruling is known as the **ally doctrine.**

Still another one of a labor union's objectives during a strike might be to induce the union employees of other contractors and subcontractors on the site to respect their picket line by declining to work. This type of picketing has the purpose of inducing a secondary boycott but this type of secondary boycott is not prohibited by the NLRA. The objective of all picketing at a jobsite is to influence secondary employees and employers to withhold their services or trade. The courts have reasoned that primary picketing that provokes secondary employees to respect a picket line is not the same thing as a secondary boycott that induces secondary employers to cease doing business with a primary contractor (or that induce secondary employees to force secondary employers to do so).

Common Situs Picketing

On any construction site you will find one or more contractors and a multitude of subcontractors, all working on the same project at the same time. They are interdependent. What affects one contractor impacts every other contractor. It stands to reason that any picket line set up at a construction site is going to impact all contractors, subcontractors, and trades people alike. It is entirely unrealistic to think otherwise. The consequence of this reality is that when picketing occurs it becomes difficult, if not impossible, to differentiate between a legal primary or secondary boycott and an illegal secondary boycott.

The courts have found that the practical solution to this dilemma is to prohibit common situs picketing. **Common situs picketing** is the picketing of an entire construction site. When a union announces a strike, a general contractor who establishes two or more gates to a site may designate separate gates, one or more gates designated for picketing, and one or more gates where picketing is prohibited. This provides secondary employees the opportunity to enter the construction site free of intimidation from picketers.

Jurisdictional Picketing

Within the construction trades, jurisdiction refers to the work performed by a particular trade and the union rules and their authority to govern that work. Jurisdiction is a local matter. Union trades in a particular geographical region are expected to honor the jurisdiction rights of all other union trades.

Jurisdictional picketing might arise in a dispute between two different trades unions who claim jurisdiction over the same type of work. Jurisdictional picketing is prohibited if one of the trades was certified for that work by an NLRB order (NLRA, §508(b)(4)(D), 1935). If a jurisdictional claim is filed with the NLRB, the unions are given 10 days to settle. If they cannot settle the NLRB assigns the work to one of the trades. The trade to whom the work is assigned may picket if the other trade does not honor that decision.

Hot Cargo Clauses

A **hot cargo clause** is a stipulation in a labor agreement that authorizes employees to decline to accept the work product of any employer in a labor dispute. Although hot cargo clauses are generally banned by the NLRA, an exception is drawn for construction work. Construction industry collective-bargaining agreements may include hot cargo clauses, "...relating to the contracting or subcontracting of work to be done at the site of the construction, alteration, painting, or repair of a building, structure, or other work..." (NLRA, §158(3), 1935).

18.5.3 Remedies

The NLRB is required to seek an immediate injunction against picketing when a complaint alleging a violation of NLRA prohibitions is asserted for secondary picketing under §158(b)(4), recognitional picketing under §158(b)(7), or hot cargo agreements under §158(e) (NLRA, 1935). If the NLRB holds the conduct illegal, it will issue a cease-and-desist order. Either a primary or secondary employer may sue in federal court to recover civil damages caused by violations of NLRA §158(b)(4), regardless of whether an unfair labor practice charge has been filed with the NLRB (NLRA, §160, 1935).

National Emergencies

The Taft-Hartley Act charged the NLRB to respond when a strike or threatened strike poses a threat to national health or safety. The U.S. President is authorized to appoint a board of inquiry to report on the issues of the dispute. The U.S. Attorney General can secure an injunction to forestall the strike for up to eighty days. The Federal Mediation and Conciliation Service (FMCS) attempts to resolve the dispute but the parties are not bound by the recommendations. If the employees will not accept the employer's last offer, the President may refer the issue to Congress for "appropriate action". The emergency provisions allow the President to delay a strike but not to address the causes of the strike.

18.6 Labor Agreements

A **union security clause**, also known as a **pre-hire agreement**, is a universal feature of collective-bargaining agreements. The union security clause commits an employer to cause its new hires to join the union, in which case its workplace is known as a **union shop**, or to either join the union or pay union dues if the new hire declines to join the union, in which case its workplace is known as an **agency shop**.

A union shop differs from a closed shop in that a union shop requires employees to join the union after being hired whereas a **closed shop** can only hire employees who were union members before being hired. The Labor Management Relations Act of 1947 outlawed closed shops (Labor Management Relations Act, 1947). The rationale for an agency shop is that because all workers, whether union members or not, enjoy the benefit of the collective bargaining agreements, all workers should share the burden of the union dues that make those benefits possible.

Some states have enacted legislation that prohibits union security clauses. These states are known collectively as **right to work states**. As of 2013, right to work was statutory law in Michigan, Indiana, Tennessee, Virginia, North Carolina, South Carolina, Georgia, Florida, Alabama, Mississippi, Louisiana, Texas, Arkansas, Oklahoma, Kansas, Nebraska, Iowa, South Dakota, North Dakota, Wyoming, Idaho, Utah, Arizona, and Nevada. In right to work states, new hires are free to join the union or not. These workplaces are known as **open shops**. Non-joiners get the benefits of the collective-bargaining agreement without having the pay the dues that make those benefits possible. Many employers in right-to-work states do not have collective-bargaining agreements. These workplaces are known as **merit shops**.

Union governance is primarily entrusted to a local union leader. A **union local** typically has jurisdiction over the union members of a particular craft in a prescribed geographic region. A **business agent**, elected by local union membership, represents the local union. Business agents are charged with promoting the best interests of the union. They wield considerable power. Some, but not all crafts, will designate a job steward. A **job steward** is the union representative for a particular trade on a construction site. Some locals will assign the job steward's position to the 1st tradesperson working on site. Other locals will allow the job steward to be appointed either by the business agent or by majority vote of the members of the local.

Many union locals will establish hiring halls. A **hiring hall** is a legal job-referral mechanism if: it does not discriminate against non-union workers; an employer can reject any referred applicant; and the hiring hall posts notice of nondiscriminatory operations. Unions may set membership and referral skill levels. Craft unions in larger metropolitan areas have established training and apprenticeship programs. Local contractors covet the highly skilled tradespeople that these programs produce. The availability of a pool is highly skilled tradespeople at the local hiring hall is one of the principal benefits of contracting in union environments.

Some craft unions have established agreements with large, national employers or employer groups that are effective wherever work is performed throughout the nation. These **national agreements** take precedence over any local agreements wherever the employer is a signatory to the national agreement. Large individual projects may utilize a project labor agreement. A **project labor agreement** is an agreement that governs all labor unions for the duration of a project. It solves all same-site union and non-union wages, benefits and jurisdiction issues. Projects benefit from consistent work rules and practices.

Chapter 19 EMPLOYMENT LAW

19.1 Introduction

As if managing the risks, schedule, cost and quality of a complex construction project isn't enough to keep him or her fully occupied, a project manager is usually accountable for human resource management at the construction site too. Contractors need a legal compliance strategy so that they can comply with the law, institute policies that prevent employment law violations, avoid legal problems, determine how issues are to be handled, and decide when to settle or to litigate claims. Project managers need to understand the basics of employment law so that they can recognize situations that raise legal concerns, make informed field judgments, and know when to call the lawyers. Noncompliance with the law is not an option.

Background

U.S. employment law ascends from a patchwork of federal, state, and local statutory law; federal agency regulations; common law judicial decisions and constitutional protections. The 14th Amendment of the U.S. Constitution, the Civil Rights Act of 1964, the Age Discrimination in Employment Act, the Equal Pay Act, and the Americans with Disabilities Act promote equal employment opportunity and prohibit discrimination in employment. The National Labor Relations Act and the Labor Management Relations Act regulate union organizing activities and secure rights to collective bargaining. The Fair Labor Standards Act, the Occupational Safety and Health Act, and the Family and Medical Leave Act specify minimum standards of pay, promote workplace safety, and otherwise shape the contours of the employment relationship. The 4th Amendment of the U.S. Constitution and the Fair Credit Reporting Act assert fundamental rights then extend and apply them to right commercially induced wrongs. Victims of negligence, defamation, or emotional distress inflicted upon them through their employment can seek redress through tort claims and breaches of the employment contract. These torts can be remedied through contract claims.

Most employment laws have come into being within the last fifty years and these employment laws continue to evolve (Walsh, 2010, pp. 3-4, 27-28). Many of our employment laws reflect small business policies, equal employment opportunity, affirmative action, and other socioeconomic policies. The history of employment law is the written narrative of ever-evolving U.S. societal and cultural values (See Figure 19.1).

The Progressive Movement	Workers' Compensation Statutes
The New Deal/Depression	National Labor Relations Act
Civil Rights Movement	Civil Rights Act of 1964 Age Discrimination in Employment Act Equal Pay Act
1970s-1980s	Common Law Claims
1980s thru current	Benefits (health insurance, pensions)

Figure 19.1 Employment Law Evolution

Employment-at-Will

Ironically, we do not have any common law right to be employed or to retain our employment. This doctrine, known as **employment-at-will**, asserts that the employment relationship can be severed at any time and it can be severed for any reason not specifically prohibited by law. Employers may terminate employment for cause, out of ignorance, or for no reason whatsoever, and do it with impunity as long as the termination is not founded in actions protected by employment discrimination laws, public policy, or private contract. Most terminations are lawful without regard to the reason for termination.

When an **employment contract** is executed between employee and employer it creates an exception to the employment-at-will doctrine. Neither employee nor employer may terminate an employment contract without sufficient cause that rises to the level of material breach. To be enforceable, however, an employment contract must be written; it must express a definite salary; it must cover a specific period of time; and both parties must sign it. Commitments for employment in construction and construction management rarely qualify as employment contracts.

An offer of employment is not an employment contract either, even if it specifies that it is an offer for permanent employment. Moreover, an offer of employment can be withdrawn, without penalty at law, at any time. And once accepted, the employment-at-will doctrine asserts that employment can be terminated for any reason not specifically prohibited by law.

Yet even when an enforceable employment contract exists, the viable remedies to cure a breach are limited. Specific performance, a court order to resume employment, is contrary to public policy against forced indentureships. Damages that flow from the breach of an employment contract may be hard to assess. A wrongfully terminated employee's damages could consist of lost income but that loss must be offset by any new employment that the employee finds and the employee will be required to mitigate damages by seeking new employment. Conversely, when an employee wrongfully breaks an employment contract the employer's damages may be limited to the cost of recruiting a substitute employee. In the final analysis, a valid employment contract does not provide much legal protection to either party.

Public sector employees have unique status. Making up roughly 15 percent of the workforce, public sector employees enjoy constitutional protections over their employment. Moreover, state or municipal civil service laws and tenure provisions often provide them with additional statutory protections. However, public sector employees are subject to restrictions on the political activities that they may engage in and they have limited ability to sue for violations of federal law. Contractors who do business with public agencies and who meet certain other criteria are required to implement small business policies, equal employment opportunity, affirmative action, and other socioeconomic policies as a condition of their contracts (Kelleher, Abernathy and Bell, 2008).

Union workers have many of their terms and conditions of employment spelled out in enforceable **labor agreements**. Employers in unionized workplaces are also more limited in their ability to make changes in workplace practices without first negotiating those changes with unions. In contrast to at-will employment of most non-union workers, discipline or discharge of a unionized employee is contractually limited to situations where the employer can establish just cause.

An employee's rights are affected by where he or she happens to live. States are free to enact laws pertaining to issues not addressed by federal law. State laws are important not only because they reach smaller workplaces than federal employment laws, but also because they sometime provide employees with rights not available under federal law. Examples include prohibitions against discrimination based on sexual orientation, higher min-

19.2 Employment Discrimination

Laws prohibiting discrimination comprise a large and absolutely central part of employment law. In 1964, the major characteristics of discrimination were overt exclusion and segregation. Workplace demographics, cultural shifts and societal attitudes have changed since that time. Other forms of discrimination now loom large, including harassment, obstacles to advancement, pay inequities, and retaliation. Discrimination increasingly takes the subtle form of blocked access to better jobs rather than outright exclusion from workplaces (Lawton, 2000). As employees become mindful of their rights and increasingly apt to challenge discriminatory practices, retaliation claims have become more common.

Evidence of discrimination in today's workplace include: 1) noticeably different employment outcomes such as concentration in lower-level jobs, higher unemployment rates, and pay inconsistencies across different races, sexes, and national origin groups; 2) increasing numbers of discrimination cases heard by administrative agencies and courts; and, 3) extensive evidence of discrimination found when employment testers are sent out to test for discriminatory hiring practices (Walsh, 2010, pp. 60-61). Despite this evidence, specific instances of discrimination remain difficult to prove.

Federally protected employment classes include race, gender, national origin, religion, citizenship, age (40 and over), disability, pregnancy, military service, and genetic information. Protected classes recognized under some state laws but not federal laws include marital status, sexual orientation, tobacco/alcohol use, HIV/AIDS, arrest and criminal convictions, weight, and personal appearance.

Employment discrimination is the unlawful limitation or denial of employment opportunity. Unequal and unfair treatment of individuals is insufficient to establish unlawful discrimination. To sustain a discrimination charge, biased treatment must arise from or relate to **protected class** characteristics. Protected class characteristics are the distinguishing markers of people who are members of protected classes. Fixed and unchangeable markers such as skin color and genetic profiles are known to have little or no relationship to the ability to do a job. Reliance upon these markers is socially objectionable because they affirm long-standing forms of hatred and prejudice. Protected class characteristics should never be used as grounds for making employment decisions.

To comply with antidiscrimination laws requires consistent, evenhanded treatment of employees. Employers that strive to be fair and to treat like situations alike regardless of the employees involved will be far less likely to discriminate. Although employers need to exhibit consistency and evenhanded treatment, they also need to be accommodating to the particular needs of disabled employees and employees whose religious practices conflict with workplace requirements. Employers need to exercise particular care to avoid antagonistic actions against employees who have filed charges or spoken out about discrimination.

Noncompliance with antidiscrimination laws can be depicted as either:

1. Disparate Treatment;
2. Adverse Impact;
3. Failure to Reasonably Accommodate; or
4. Retaliation.

To establish that discrimination has occurred a causal link must be shown between the employer's actions and the protected class characteristics of the employee.

Disparate Treatment

Disparate treatment occurs when there is disparity or inequality in how employees are treated, and the difference in treatment is due to the employee's race, gender, or other protected class characteristic. The key element of disparate treatment is discriminatory intent. The meaning of intent is that the decision maker based his or her decision, in whole or part, on a protected class characteristic of the affected employee. Disparate treatment cases exhibit different employment behavior patterns. These patterns are discernable as:

- Direct Evidence,
- Facially Discriminatory Policies or Practices,
- Reverse Discrimination,
- Pretext,
- Mixed Motives,
- Pattern or Practice, or
- Harassment.

Direct Evidence

Direct evidence consists of any clear and unambiguous assertion by a person authorized to make an employment decision that his or her adverse employment decision was based upon the protected class characteristics of the affected employee. The assertion may be either spoken or written but it must be made reasonably close in time to the adverse employment decision.

Facially Discriminatory Policies or Practices

The Civil Rights Act of 1964 permits discrimination on the basis of

> "...religion, sex, or national origin in those instances where religion, sex, or national origin is a bona fide occupational qualification reasonably necessary to the normal operation of the particular business or enterprise..." (Civil Rights Act of 1964 [Civil Rights Act], §2000e-2(e), 1964).

Facially discriminatory policies or practices involve direct evidence of discrimination that an employer admits to while contending that the adverse employment decision was necessitated by valid business objectives. To avert charges of facially discriminatory policies or practices an employer must prove that a **bona fide occupational qualification [BFOQ]** existed. The BFOQ must be shown to be "reasonably necessary to the normal operation of the business..." and therefore it justified lawful discrimination based upon the protected class characteristics of the affected employee.

For example, a Catholic school may choose to only hire Catholic school teachers because they promote Catholicism, or a women's clothing designer may hire only female models who are 5'7' and 114 pounds because that is the size you must be to fit into their clothing, or a casting director hires only an African American woman in her sixties because that is what the role calls for. These are all BFOQs. In contrast, white Christian male around 6'2" and 180 lbs. is not a BFOQ for a construction manager.

Reverse Discrimination

Reverse discrimination claims arise when laws and policies, such as affirmative action plans, result in employment practices that favor a protected class and a qualified member of a non-protected class claims that he or she was denied an opportunity as a direct consequence of favoring the protected class. Whether or not reverse discrimination claims are sustained depends on whether or not the employer's preferential treatment actions are lawful or not.

Pretext

Pretext is a ruse to cover up an employer's discriminatory employment practices. Pretext is engaged when an employer asserts that its motivation for an adverse employment decision was something other than a decision based upon the protected class characteristics of the affected employee. Pretext might be found in statements that a chosen candidate who was not members of a protected class was more qualified, promotable, or had more experience. To steer clear of pretext, it behooves employers to insert clear, unambiguous, and measurable criteria in their job descriptions; and to apply them rigorously and equally to all candidates for employment, promotion or advancement.

Mixed Motive

Mixed motive cases are characterized by adverse employment decisions that are motivated by multiple factors, one or more that unlawfully discriminatory and others that are lawful. If only one of the factors contributing to an adverse employment decision is found to be unlawful, the employer is guilty of employment discrimination. Damage award in mixed motive cases may be reduced in consideration of any lawful motives, particularly if it is found that the lawful motive by itself was enough to cause the adverse employment decision that was made.

Pattern or Practice

Pattern or practice cases arise from systematic discriminatory practices by an employer that affects groups of employees over periods of time. Typical pattern and practice cases complain of the segregation of employees in protected classes into lower paying, less desirable jobs. An employer's historical records and statistical analysis tools are utilized to demonstrate a pattern or practice of discrimination in the workplace.

Harassment

Harassment is mistreatment of an employee or groups of employees that is founded on their protected class characteristics. Sexual harassment is the most common and pervasive form of harassment. Typically, accusations of sexual harassment claim requests for sexual favors, inappropriate verbal or physical sexual conduct, or unwelcome sexual advances, frequently perpetrated by an employee's supervisor or other managers with authority to make employment decisions. Harassment can also be construed when an employee is subjected to inferior working conditions because of the employee's gender.

Harassment must occur on the job. Conduct occurring off the job is not relevant in these cases. The magnitude of an employer's liability for harassment depends upon the nature and severity of the harassment and the harasser's position of authority in the company. Employers are strictly liable for harassment by its executive managers, regardless of the nature and severity of that harassment.

Adverse Impact

Adverse impact, also known as **disparate impact**, is any adverse employment effect suffered by a protected class group due to a business practice or requirement that is not job related and consistent with business necessity. An employer may insist that these business practices or requirements are neutral and that they are not intended to have an adverse effect on any protected class group. It is not, however, necessary to make a showing of an employer's intent in adverse impact cases. It is only necessary to show a factually adverse effect on the protected class group. As long as other business practices or requirements exist that would not produce the same adverse impact, and the employer failed to use them, adverse impact discrimination will be sustained against the employer.

Examples of neutral requirements or practices that have been shown to have adverse impact include employment tests, height and weight requirements, language requirements, physical strength tests, military service, type of military discharge, arrest and conviction records, and educational requirements (Walsh, 2010, pp. 81-85). Whether or not any of these neutral requirements or practices would sustain adverse impact discrimination in any particular case depends on the nature and circumstances of their use.

Failure to Reasonably Accommodate

Employment law requires employers to provide reasonable accommodations when needed by either disabled employees or employees whose religious beliefs and practices conflict with their employer's rules and policies. As long as it does not expose the employer to an undue hardship, an employer must find a way to accommodate these employees. Employers need to be flexible and accommodative in their hiring practices, job performance evaluations, work schedules, and other rules, workplace practices, and expectations in consideration of their employee's disabilities and religious beliefs. Failing to do so, a charge of failure to reasonably accommodate can be sustained against the employer.

Retaliation

Retaliation is any act of retribution against an employee for his or her involvement in protected civil rights activities. Retaliation cases have four elements: 1) the employee's actions arose from protected activities; 2) those actions were not illegal; 3) the employer's reaction was materially adverse; and 4) there must be a causal connection between the employee's actions and the employer's reaction. All four elements must be proved for a retaliation charge to stand.

There are two broadly defined classes of protected civil rights activities: participation and opposition. **Participation** refers to the enforcement of antidiscrimination laws. It involves activities such as talking to investigators, filing a lawsuit, or testifying in court. **Opposition** refers to supporting civil rights through activities other than participation. Examples include complaining about discrimination in an internet blog, making accusations of discrimination to company managers, or talking to civil rights organizations. Employees who participate in enforcement or oppose discrimination in these ways are navigating in the safe harbor of protected activities.

The protected activity safe harbor is breached, however, when the activity is unlawful. It is lawful for employees to post company information to the internet but is unlawful for employees to post confidential company information to the internet. It is lawful to acquire information but it is illegal to trespass to acquire that information. It is lawful to publish statements about a company but it is libelous to publish false and defamatory statements about a company. Civil wrongs (torts) and criminal acts are not going to be protected, even when they were motivated by the highest moral purposes.

The employer's reaction to an employee's action must be materially adverse. In other words, the employer must have actually done something harmful to the employee. Claims that the employer threatened to retaliate are not sufficient. The employer must have actually retaliated. The retaliation, however, can be much less severe than the firing or demotion of the employee. Both job reassignment and suspensions have been deemed materially adverse even when they did not result in the loss of wages.

Finally, there is the element of causal connection. It must be shown that but for the employee's protected activities the employer would not have punished the employee. Employers are allowed to terminate employees, demote them, reassign them or otherwise punish them for good reason or for no reason whatsoever in an employment-at-will setting. If an employer can show that the punishment was the direct consequence of something else, not a protected activity, then the employer will be excused.

19.2.1 Equal Pay Act

Pay discrimination based on gender is a particular concern. A substantial gap still exists between the earnings of full-time male workers and full-time female workers. A methodology for proving pay discrimination has been developed under the **Equal Pay Act**. The Equal Pay Act of 1963 prohibits discrimination in pay based on gender. The Act has been codified into the FLSA as §206(d), in pertinent part as follows:

"(d) Prohibition of sex discrimination
  (1) No employer...shall discriminate...between employees on the basis of sex by paying wages to employees...at a rate less than the rate at which he pays wages to employees of the opposite sex...for equal work on jobs the performance of which requires equal skill, effort, and responsibility, and which are performed under similar working conditions, except where such payment is made pursuant to (i) a seniority system; (ii) a merit system; (iii) a system which measures earnings by quantity or quality of production; or (iv) a differential based on any other factor other than sex." (FLSA, §206(d), 1935).

Four factors are used to determine whether jobs are substantially equal: skill, effort, responsibility, and working conditions. Skill is marked by effective use of knowledge, superior performance, sound execution, physical dexterity, competence and ability. Effort is the physical exertion and mental tenacity required. Responsibility is the moral, legal and mental accountability, reliability and trustworthiness required. Working conditions allude to constraints such as hazardous physical environments, poor air quality and harsh weather.

To sustain a case of pay discrimination an aggrieved employee needs to identify comparable employment in the same company, by a person of the opposite gender, earning a higher wages for doing substantially the same work. An employer can defeat the claim by demonstrating that the difference in wages was attributable to a seniority system, a merit system, a system that measures earnings by quantity or quality of production, or a differential based on any other factor other than sex.

Eradication of pay discrimination based on gender has proven to be an elusive national policy goal. Despite that over 50 years have elapsed since passage of the Equal Pay Act, a gap still exists between the earnings of male and female workers.

19.3 Fair Labor Standards

The principal federal statute regulating wages and hours is the **Fair Labor Standards Act [FLSA]**. The FLSA's main requirements are straightforward. This law establishes a federal minimum wage and requires premium pay for overtime work. It also sets out certain work-hour limitations for minors (FLSA, 1939).

FLSA requirements do not apply to all employees. The terms **exempt** and **nonexempt** are used to contrast employees for whom employers do have to follow FLSA requirements and those for whom they do not. Employers must pay nonexempt employees an hourly rate at least equal to a legislatively determined minimum wage for each hour worked in a workweek, premium pay for overtime work, and the work hours of minors are restricted.

The federal **minimum wage** as of July 24, 2009 is $7.25/hour (FLSA, §206(a)(1)(C). An exception to hourly minimum wage requirement is the **opportunity wage**. Employees under 20 years of age can be paid by employers at the rate of $4.25.hour for their first ninety calendar days on the job (FLSA, §206(g)(1), 1939). The requirement applies to gross pay. Gross pay is the amount of money earned before withholding of income taxes, and the employee's share of social security, Medicare taxes and deduction of other benefits. Annual or monthly salaries or piecework must be converted to an hourly rate in order to ascertain that the minimum wage is being met.

The basic unit of time for determining compliance with the FLSA's requirements is the workweek. The workweek is any fixed and recurring period of seven consecutive days (168 hours). The FLSA does not limit the number of hours employees can be required to work in any workweek, but employers must pay at least one and one-half times an employee's regular rate of pay for each hour worked in excess of forty in a workweek (FLSA, §207(a), 1939).

Whether a company may pay for overtime work with **compensatory time off** rather than with overtime pay depends on whether the employer is in the private or public sector. Private employers are not allowed pay for overtime required under the FLSA with compensatory time off, whereas government agencies are allowed to do so (FLSA, §207(o), 2014).

Some states have different minimum wage laws. Employers must pay the higher of the federal minimum wage rate or the state minimum wage rate. The California minimum wage as of July 1, 2014 is $9.00/hour and $10.00/hour on or after July 1, 2016 (California Labor Code, 2014).

The White Collar Exemption

Employers do not have to pay either the minimum wage or premium pay for overtime work to employees who are exempt. This exemption, the so-called "white collar exemption" applies to people employed as executives, administrators, or professionals; computer systems analysts, computer programmers, software engineers, or other similarly skilled workers; outside salespersons; or highly compensated employees (FLSA, §213(a), 1935). It is not the job title that determines exempt status though. It is the duties that the employee performs and the salary earned for performing those duties.

The primary duties of executives are managing a business, or one of its subdivisions or departments; directing the work of two or more other employees; authorizing the hiring or firing of other employees; or making suggestions and recommendations as to the hiring, firing, advancement, promotion or any other change of status of other employees.

The primary duties of administrators are performance of office or non-manual work directly related to the management policies or general business operations of the company and the exercise of discretion and independent judgment with respect to significant business matters.

The primary duties of professionals are the application of knowledge of an advanced type in a field of science or learning customarily acquired by a prolonged course of specialized intellectual instruction; or the application of invention, imagination, originality or talent in a recognized field of artistic or creative endeavor. The former are known as "learned professionals." The later are known as an "artistic professionals." An individual who holds an undergraduate or graduate degree in construction management, engineering, or planning as a qualification for his or her employment would generally be considered a learned professional. An individual who holds an undergraduate or graduate degree in architecture, landscape architecture or interior design as a qualification for his or her employment would generally be considered an artistic professional.

The primary duties of computer-related employees are the application of systems analysis techniques and procedures, including consulting with users to determine hardware, software or system functional applications; design, development, documentation, analysis, testing, creation or modification of computer systems or programs including prototypes, based on and related to user or system design specifications; design, documentation, testing, creation or modification of computer programs related to machine operating systems; or a combination of duties described in the above three options that requires the same level of skill to perform those duties.

The primary duties of outside salespersons are to make sales; to obtain orders or contracts for services; or to obtain contracts for the use of facilities for which clients or customers pay. Outside salespersons typically perform their duties away from their employer's place of business.

To be exempt, an executive, administrator, professional or computer-related employee must both perform the appropriate duties and earn $23,660 or more annually, or $455 or more per week (or $27.63 per hour for a computer-related employee). Most executive, administrative, professional, or computer-related employees who earn less than $23,660 annually, or less than $455 per week (or $27.63 per hour for computer-related employees) are nonexempt, regardless of the duties that they perform. An executive who owns at least a 20% equity interest in its business and is actively engaged in managing its business is exempt regardless of earnings. Outside salespersons are also exempt regardless of earnings. Also exempt are highly compensated employees - defined as employees who both earn over $100,000 annually and perform one or more of the duties of executives, administrators or professionals.

On July 1, 2015, the Department of Labor issued proposed regulations that, if adopted as proposed, would substantially increase the minimum salary needed to qualify as exempt. It is expected that this could increase the minimum salary to $50,440 annually or $970 per week.

Compensation Categories

Compensation takes many forms. Employees may receive hourly wages, tips, commissions, piecework earnings, bonuses, merit pay, lodging/meals, pay for holidays/vacations/sick days, premium pay (weekends/night shift), and profit sharing/benefits. When calculating the regular rate of pay on which overtime pay is based, most forms of compensation are included. The primary exclusions include most paid absences for vacations, holidays, and illnesses; discretionary bonuses and prizes not based on attendance or merit; reimbursements for travel expenses and material costs paid by the employee; employer payments for pensions and other employee benefits; employer expenses for profit-sharing plans; premium pay for working on holidays; and daily or other non-FLSA required overtime pay.

The amount of time actually spent performing work duties determines compensable time. An employee taking too long to carry out some task is a performance issue and not an excuse for violating the FLSA. Work activities must be sufficiently related to employees' primary job duties to be considered compensable. Typical compensable work includes employer-required training; traveling between work sites; waiting while on duty; restrictive on-call arrangements; meal periods when not substantially relieved from work duties; and rest periods of twenty minutes or less. Typical noncompensable work includes time consumed with pre-employment tests; voluntary training that is not related to regular duties, is outside work hours, and that is during time when work is not performed; traveling back and forth from home to a work site; waiting to start work; waiting after being relieved from duty for a definite and useful time; most on-call duty performed outside the workplace; and meal periods free of duties (and usually at least thirty minutes in duration).

The federal government does not limit the number of hours worked except for airline pilots, truck drivers and child labor. Employers must ascertain the ages of their youthful employees with certainty because they will be held strictly liable for employing child labor, regardless of whether the youth presented false age credentials.

19.4 Occupational Safety and Health

Providing safe workplaces may be the single most important responsibility that a contractor has to its workers. Regrettably, the safety performance of the U.S. construction industry lags behind that of other industries. Construction remains a hazardous occupation and liability for construction injuries and death generally attach to a contractor.

The Occupational Safety and Health Act

The most pervasive regulatory scheme attached to construction projects is that which flows from the **Occupational Safety and Health Act of 1970 [OSH Act]**. The OSH Act asserts, "...Each employer . . . shall furnish to each of his employees employment and a place of employment which are free from recognized hazards that are causing or are likely to cause death or serious physical harm to his employees..." (Occupational Safety and Health Act of 1970 [OSH Act], §654(a)(1), 1970).

Three separate agencies were created with the OSH Act. The **Occupational Safety and Health Administration [OSHA]** administers the OSH Act, establishes standards, provides information about how to meet those standards, and enforces them by conducting inspections. The **National Institute of Occupational Safety and Health [NIOSH]** provides scientific and technical advice to OSHA. The **Occupational Safety and Health Review Commission [OSHRC]** hears appeals of OSHA enforcement decisions.

OSHA covers private sector employers and employees in all 50 states, the District of Columbia, and other U.S. jurisdictions either directly or through an OSHA-approved state program. State-run health and safety programs may supplant Federal OSHA but they must demonstrate that they are at least as effective as the Federal OSHA program.

The United States Postal Service (USPS) is covered by OSHA. Other Federal agencies are not, but they must all have a safety and health program that meets the same standards. And although OSHA cannot fine federal agencies, it does monitor federal agencies and responds to workers' complaints. Employees who work for state and local governments are not covered by Federal OSHA, but have OSH Act protections if they work in a state that has an OSHA-approved state program. The states of Connecticut, Illinois, New Jersey, New York, and the Virgin Islands territory have OSHA approved plans for their public sector employees.

OSHA Standards

OSHA endeavors to define minimum levels of safety in two ways: 1) through **OSHA standards** that address specific hazards; and, 2) through the OSH Act's general duty clause. OSHA creates both general industry safety standards that apply to all industries and specific safety standards for the maritime, agriculture and construction industries. Employers must comply with all standards that apply to their operations. In deciding what standards are applicable, more specific standards take precedence over more general ones.

The **general duty clause** is the fallback position for hazards that might not be addressed by OSHA standards. The general duty clause places responsibility for workplace safety on the employer. OSHA will enforce that responsibility by invoking the general duty clause in the absence of specific standards.

OSHA Inspections

OSHA inspectors will examine a worksite in response to employee complaints of safety violations. They will also conduct unannounced inspections to determine whether employers are complying with the law. Most OSHA inspections are unannounced. OSHA inspectors are required to enter workplaces at reasonable times to examine records, inspect conditions, and question individuals. Inspections are prioritized toward what appears to be the most imminent dangers. Dangerous projects will be set up with programmed inspections

A typical inspection commences with an opening conference facilitated by the OSHA compliance officer. Following that there is a jobsite tour. Both employer representatives and union representatives might accompany the compliance officer as the jobsite tour progresses. The compliance officer points out any OSHA violations. Some violations may be minor enough to correct on the spot. More serious violations will result in a **citation**. At the end of the inspection there will be a closing conference at which time the compliance officer will state any violations and explain the appeal process

After review of the compliance officer's inspection reports, the director of the OSHA area office will send notice to the contractor indicating the nature of the violations, the OSHA standard(s) violated, the monetary penalties associated with the violations, and the length of time that the employer has to correct the problems. This correction time is known as the **abatement period**.

Fines vary according to how severe the violations are and whether they are willful or a death was involved. The maximum penalty is a $20,000 fine and one year in jail, regardless of the number of violations. OSHA enhances the impact of its enforcement effort by cooperating with other agencies, such as the EPA.

Copies of citations received must be posted in the workplace near where the violations occurred. Postings must remain in place for three working days or until the violations are corrected, whichever is longer. Employers have fifteen days following the receipt of a citation to appeal to the OSHRC, to petition OSHA to modify abatement schedules, or to meet with OSHA area directors to discuss settlements

OSHA relies on employee reporting of potential safety hazards. OSHA is prohibited from revealing the identities of employees who make safety complaints. It does, however, inform employers when inspections are prompted by complaints and provides copies of written complaints with the complainants' names deleted. Employers are prohibited from retaliating against employees who report potential safety hazards.

Recordkeeping

The OSH Act requires employees to keep records of occupational injuries and illnesses. This recordkeeping includes:

- Communicating specific procedures for reporting workplace injuries and illnesses;
- Recording (within six days of occurrence) all work-related injuries or illnesses that relate to death, days away from work, restricted work, transfer to another job, loss of consciousness, or medical treatment beyond first aid;
- Maintaining records for each separate establishment and not just an entire company;
- Posting an annual summary of injuries and illnesses for employee inspection during the month of February (for the previous year);
- Retaining records for documenting safety training provided and certifying that certain potentially dangerous equipment has been examined and is in safe working order;
- Saving records from any periodic medical tests screening or monitoring for adverse health effects;
- Retaining medical and exposure records for many years since occupational diseases often take a long time to develop; and

- Reporting to OSHA (within eight hours) any fatal accident or any accident that results in the hospitalization of three or more employees (Walsh, 2010, pp. 491-492).

Employers are also required, if requested, to participate in the Bureau of Labor Statistics' (BLS) annual survey of occupational injuries and illnesses, a basic source of data for occupational safety and health statistics.

Workplace Injuries

There are several things an employer should do when an injury occurs in its workplace. They include:

- Requiring that employees report all injuries that occur in the workplace as soon as possible;
- Erring of the side of caution in referring injured employees for medical treatment;
- Investigating reports of injuries immediately and thoroughly;
- Identifying (and abating) hazards that might have caused an injury;
- Not retaliating against employees for filing workers' compensation claims;
- Staying in close contact with injured employees and their medical care providers;
- Assigning light-duty work (or making it available) for the injured employees who are not yet ready (or capable) of fulfilling all the duties of their former positions; and
- Being mindful that employees with work-related injuries might also be entitled to take leave or receive accommodations that will allow them to perform the essential functions of their regular jobs (Walsh, 2010, pp. 507-508).

19.5 Enforcement

With few exceptions public agencies charged with enforcing employment laws rely on workers' claims of employment law violations. OSHA is an exception because they will often initiate a safety inspection. But all other agencies wait for workers to come forward before bringing and resolving charges. Procedures for bringing complaints forward vary, depending on the law that forms the basis of the complaint and the particular agency charged with regulating pursuant to that law.

Workers need to file timely complaints because the length of time available for them to file a complaint is constrained by a **limitation period**. A worker's right to file is forfeited if the filing is not made within the relevant limitation period. These limitation periods vary from agency to agency. When the complaint covers repeated acts upon an aggrieved employee the tolling of the limitation period begins on the last such act, but the complaint attaches to all such acts, even those that were far distant in time. It behooves employers to retain employment files almost indefinitely, in order to defend actions taken in the distant past with respect to workers who might not even work for their company any longer.

Workers are given different means to enforce their rights against employers. Most employment law claims can be filed in state or federal courts. Key exceptions to this rule are OSHA safety hazard violations, labor disputes, discrimination cases and wage-and-hour violations.

OSHA complaints precipitate a site inspection by an OSHA compliance officer and a follow-up by the director of the OSHA area office. Labor disputes are adjudicated by the NLRB, whose General Counsel may file a compliant in a federal court. Discrimination cases and wage-and-hour violations must be brought before the Equal Employment Opportunity Commission [EEOC] of the Wage and Hour Division of the Department of Labor.

If the EEOC rules affirmatively on the merits of the case it will issue a **right to sue letter**, allowing the aggrieved employee to file a suit for damages in state or federal court. Should the EEOC rule for the employer instead, they will not issue a right to sue letter and the aggrieved employee is barred from bring suit in any court. A small percentage of cases are brought directly to court by the EEOC.

Any lawsuit that an individual can bring to court can also be brought by numerous plaintiffs through a **class-action lawsuit.** To bring a class action lawsuit the plaintiffs must first have their class certified by the court, by demonstrating that the same employer violated the rights of every member of their class in a substantially similar manner. If the class action is sustained, the class members share any damages awarded by the court. Class-action lawsuits are efficient ways of dispensing justice in large cases.

Arbitration Agreements

Arbitration agreements might be embedded in an employment contract. Arbitration may also be instituted by collective-bargaining agreements. The Federal Arbitration Act, enacted by Congress in 1925, requires courts to enforce most written arbitration agreements. A valid arbitration agreement effectively eliminates a party's ability to pursue a claim in the courts, even if the arbitration agreement was compelled by the employer through a take-or-leave it contract of adhesion. All other elements of claims pursuit remain the same when there is an arbitration agreement, such as the requirement of an EEOC right-to-sue letter

The courts have sometimes refused to enforce mandatory arbitration agreements because they are found unconscionable. It takes more than unfairness to create **unconscionability**. To be unconscionable a mandatory arbitration agreement must exhibit both procedural and substantive unconscionability. **Procedural unconscionability** exists when the weaker party was prevented from seeing the unfair contract terms before signing the agreement. **Substantive unconscionability** exists when the agreement was actually and substantially unfair.

Agreements are unconscionable with a showing that:
1. A contract of adhesion was drafted by a more powerful party;
2. The contract's terms unreasonably favor the more powerful party; and
3. The weaker party was not given the opportunity to see the unconscionable elements before reaching agreement.

California courts have been especially likely to closely scrutinize mandatory arbitration agreements and find them to be unconscionable. One area of particular concern is the procedure for selecting an arbitrator. An essential requirement for a fair arbitration is neutrality. Employers who attempt to "stack the deck" risk charges of unconscionability.

REFERENCES

American Arbitration Association [AAA]. (2009). *Construction Industry Arbitration Rules and Mediation Procedures*. New York: author

American Institute of Architects [AIA]. (2000). *Document E352™-2000 Duties, Responsibilities, and Limitations of Authority of the Architect's Project Representative.* Washington, D.C.: author

American Institute of Architects [AIA]. (2007a). *Document A101™-2007 Standard Form of Agreement Between Owner and Contractor (Stipulated Sum).* Washington, D.C.: author

American Institute of Architects [AIA]. (2007b). *Document A201™ - 2007 General Conditions of the Contract for Construction.* Washington, D.C.: author

American Institute of Architects [AIA]. (2007c). *Document E201™-2007 Digital Data Protocol Exhibit.* Washington, D.C.: author

American Law Institute. (2011). *The Uniform Commercial Code*. Philadelphia: Author.

American Subcontractors Association, Inc. (May 23, 2013). *Statement for the Record, Hearing on building America: challenges for small construction contractors before subcommittee on contracting and workforce committee on small business, U.S. House of Representatives.* Alexandria, VA: author.

Bauman, J.A., York, K.H., & Bauman, J.H. (2003). *Remedies*. 11th ed. Chicago: Gilbert Law Summaries; Thomson

Calamari, J.D. & Perillo, J.M. (1999). *Contracts.* (4th ed.). St. Paul: Black Letter Outlines; West

California Environmental Protection Agency. (n.d.) *Air Resources Board.* Retrieved Aug. 13, 2001 from http://www.arb.ca.gov/homepage.htm

Castagnera, J.O. and Cihon, P.J. (2011). *Employment and Labor Law (7th Edition).* Mason, Ohio: Cengage Learning.

Child, J., Faulkner, D., and Tallman, S.B. (2005). *Cooperative strategy: Managing alliances, networks, and joint ventures.* Oxford: Oxford University Press. (as cited in Starzyk, G.F. and McDonald, M. (2011). *The Collaborative Dance; Only Three Steps.* Washington, D.C.: BIM Forum of the National Institute of Building Sciences.)

Circo, C.J., and Little, C. H., (Eds.). (2009). *A State-by-State Guide to Construction & Design Law.* 2nd ed. Chicago: American Bar Association

ConsensusDocs\™. (2011, amended 2012). *ConsensusDocs™ 200 Standard Agreement and General Conditions Between Owner and Constructor (Lump Sum Price).* Arlington, VA: author.

Contractor's Licensing Service, Inc. (n.d.). *How to Get a Contractor's State License.* Retrieved August 12, 2014 from www.clsi.com/contractor_license.htm. Albuquerque, N.M.: author.

Contractors State License Board [CSLB]. (n.d.) Retrieved August 12, 2014 from www.cslb.ca.gov. Sacramento, CA: author

Coulson, R. (1981). Preface to *Construction Arbitration: Selected Readings.* Gibbons, M. and Miller L. (Eds.)

Davies, R., Kilmann, J., Orlander, P. and Shanahan, M. (2009). *The new social contract: engaging employees during a downturn.* Retrieved July 28, 2014 from Boston Consulting Group website http://www.bcg.com/documents/file28214.pdf

Davis, T. (1993). *The Illusive Warranty of Workmanlike Performance: Constructing a Conceptual Framework.* 72 Neb. L. Rev. 981

Dobbs. D.B. (1993). *Law of Remedies: Damages-Equity-Restitution,* (2nd ed.). St. Paul: West

Eisenberg, M.A. (2002). *Contracts*, (14th ed.). Chicago: Gilbert Law Summaries

Engineers Joint Contract Document Committee [EJCDC]. (2007) *C-700 Standard General Conditions of the Construction Contract.* Alexandria, VA: National Society for Professional Engineers for EJCDC

Farnsworth, E.A. (2011). *Selections for Contracts: Restatement Second.* New York: Foundation Press; Thomson/West

Garner, Bryan A. (Ed.). (2006). *Black's Law Dictionary.* 3rd ed. St. Paul, MN: Thomson West

Goetsch, D.L. (2013). *Construction Safety and Health.* (2nd ed.). Saddle River, N.J.: Pearson Education, Inc.

Hunter, R.P. (1999). "Four Reasons for the Decrease in Union Membership". *The Mackinac Center website.* Retrieved August 14, 2013 from http://www.mackinac.org

Jurkovich, M.J. & Hebesha, A.G. (2009). Chapter 5: California Construction and Design Law. In Circo, C.J. & Little, C.H. (eds.) *A State-by-State Guide to Construction & Design Law: Current Statutes and Practices,* (2nd ed.). Chicago: American Bar Association

Kelleher, T.J. Jr. (Ed.). (3rd ed.). (2005). *Common Sense Construction Law: A Practical Guide for the Construction Professional.* New Jersey: John Wiley & Sons, Inc.

Kelleher, Thomas J, Jr., Abernathy, Thomas E. IV, and Bell, Hubert J., Jr. (Eds.). (2008). *Smith, Currie & Hancock's Federal Government Construction Contracts : A practical guide for the industry professional.* Hoboken, N.J.: John Wiley & Sons, Inc.

Kochan, Thomas A. (n.d.). *Building a new social contract at work: a call to action.* Retrieved July 28, 2014 from http://leraweb.org/sites/leraweb.org/files/Meetings/Addresses

Kolton, Kevin L. & Montgomery, Virgil R. (2007). State Regulation of the Construction Manager. In Kolton, Kevin L., Montgomery, Virgil R. & Hess, Stephen A. (Eds.). *Design Professional and Construction Manager Law.* (pp. 205, 210). Chicago: American Bar Association.

Lawton, Anne. "The Meritocracy Myth and the Illusion of Equal Employment Opportunity." *Minnesota Law Review* 85 (December 2000), 605-12.

Patrick, A.E., Beaumont, D.R., Brookie, T.L., Kirsh, H.J., Tarullo, M.D. and Spencer, K.S. (Ed's.). (2010). *The Annotated Construction Law Glossary.* Chicago: American Bar Association

Schierow, L. (April 1, 2013). *The Toxic Substances Control Act (TSCA): A Summary of the Act and Its Major Requirements.* Congressional Research Service. Retrieved Aug. 13, 2013 from http://www.fas.org/sgp/crs/misc/RL31905.pdf

Smith, Currie & Hancock, LLP, (3rd ed.). (2005). *Common Sense Construction Law: A Practical Guide for the Construction Professional.* Kelleher, T.J. Jr. ed. P. 103. New Jersey: Wiley

Staak, J.C. (2008, July/August). The gaps that bite: missing terms in contracts can come back to haunt you. *Florida Home Builder.* 26-27.

Starzyk, Gregory F. (2014). *Alliance Contracting: Enforceability of the ConsensDocs 300 Mutual Waiver of Liability in U.S. Courts.* CIB W113 Law and Dispute Resolution Working Commission at the 30th Annual Conference of the Association of Researchers in Construction Management (ARCOM). Portsmouth, UK.

Stipanowich, T.J. (1998). Reconstructing construction law: reality and reform in a transactional system. *Wisconsin Law Review.* (1998 Wis. L. Rev. 463)

Stipanowich, T.J. (Spring, 1996). Beyond Arbitration: Innovation and Evolution in the United States Construction Industry. *Wake Forest Law Review.* (31 Wake Forest L. Rev. 65)

Sweet, J. (1997). *Sweet on Construction Law.* Chicago: American Bar Association

Sweet, J. & Schneier, M.M. (2009). *Legal Aspects of Architecture, Engineering and the Construction Process.* (8th ed.). Stamford: Cengage Learning.

Sweet, J. & Schneier, M.M. (2013). *Legal Aspects of Architecture, Engineering and the Construction Process* (9th ed.). Stamford: Cengage Learning.

Sweet, J., Schneier, M.M. & Wentz, B. (2015). *Construction Law for Design Professionals, Construction Managers and Contractors.* Stamford: Cengage Learning.

The History Channel website. (2013). *Industrial Revolution.* Retrieved August 14, 2013, from http://www.history.com/topics/industrial-revolution.

The White House website. (2013). *The American Presidents.* Retrieved August 14, 2013, from http://www.whitehouse.gov/about/presidents/franklindroosevelt

Thomsen, C. and Sanders, S. (2011). *Program Management 2.0.* McLean, VA Construction Management Association of America Foundation.§

Tuchman, B.W. (1978). *A Distant Mirror: The Calamitous 14th Century* (p. 39). New York: Alfred A. Knopf.

Tuttle, C. (2010). "Child Labor During the British Industrial Revolution". *The Economic History website.* Retrieved August 14, 2013 from http://www.eh.net

U.S. Army Corps of Engineers and North Atlantic Regional Business Center. (September 5, 2008). *Integrated Design-Bid-Build and its application to the BRAC05 NGA New Campus East Engineer Proving Ground, Fort Belvoir, VA.* Retrieved July 19, 2014 from http://www.samehuntington.com/shared/content/presentations/tc_s4_jones.pdf

United States Environmental Protection Agency [EPA]. (2011a). *Protecting People and Families from Radon: A Federal Action Plan for Saving Lives.* Retrieved Aug. 13, 2013 from http://www.epa.gov/

United States Environmental Protection Agency [EPA]. (2011b). *Small Entity Compliance Guide to Renovate Right: EPA's Lead-Based Paint, Renovation, Repair, and Painting Program* (as amended in 2010 and 1011). Retrieved Aug. 13, 2013 from http://www.epa.gov/

United States Environmental Protection Agency [EPA]. (n.d.(a)). *A Citizen's Guide to Radon.* Retrieved Aug. 13, 2013 from http://www.epa.gov/radon/pubs/citguide.html

United States Environmental Protection Agency [EPA]. (n.d.(b)). *Federal Environmental Requirements for Construction.* Retrieved Aug. 13, 2013 from http://www.epa.gov

United States General Accounting Office [GAO]. (September 1999). *Fair Labor Standards Act: White-collar exemptions in the modern work place.* GAO/HEHS-99-164. p. 8. Washington, D.C.: author

Walsh, D.J. (2010). *Employment Law for Human Resource Practice* (*3rd Edition*). Mason, Ohio: Cengage Learning.

White, J.J. & Summers, R.S. (2000). *Uniform Commercial Code* (5th ed.). St. Paul: West Group

REFERENCES TO LEGAL MATERIALS

Accreditation, Certification and Work Practices for Lead-Based Paint and Lead Hazards. 17 California Code of Regulations Division 1 Chapter 8 §§35001 *et seq.* (2008). Retrieved August 15, 2014 from http://www.dir.ca.gov/dlse/ccr.htm.

Aced v. Hobbs-Sesack Plumbing Company (1961) 55 Cal.2d 573, 583 [12 Cal. Rptr.2d 257, 260]

Allied Fire & Safety Equipment Co. v. Dick Enterprises, Inc. 886 F. Supp. 491 (1995)

American & Foreign Ins. Co. v. Bolt 106 F.3d 155 (6th Cir. 1997)

American Subcontractors Association, Inc. (May 23, 2013). Statement for the Record, Hearing on building America: challenges for small construction contractors before subcommittee on contracting and workforce committee on small business, U.S. House of Representatives. Alexandria, VA: author.

Anti-Kickback Act of 1986. 41 USC §§8701-8707. (2011). Retrieved July 29, 2014 from http://uscode.house.gov.

Appeal of Pavco, Inc. ASBCA No. 23783. Armed Services Board of Contract Appeals. 80-1 B.C.A. (CCH) P14, at 412. (April 8, 1980).

Appeal of Quality Electric Service. ASBCA No. 25811. Armed Services Board of Contract Appeals. 81-2 B.C.A. (CCH) P15, 380. (August 18, 1981).

Arya Group, Inc. v. Cher, 77 Cal App. 4th 610 (2000).

Ballard and Sons v. Loving Municipal School District, 868 P.2d 646 (1994).

Beacon Homes, Inc. v. Holt, 266 N. C. 467. (1966).

Buy America Act of 1988, 41 USC §§101 *et seq.* (1988). Retrieved August 20, 2014 from http://uscode.house.gov.

C. B. Jackson & Sons Construction Company v. Robert Davis and Jean Davis, 365 So.2d 207 (1978).

California Health & Safety Code [Cal HSC] Division 13 Part 3 Chapter 9 §19827.5 Demolition Permits (1990). Retrieved August 15, 2014 from http://leginfo.legislature.ca.gov/faces/codes.xhtml.

California Civil Code [Cal CIV] Division 2 Part 2 Title 7 Chapter 2 §§896 *et seq.* Actionable Defects. (2002). Retrieved July 27, 2014 from http://leginfo.legislature.ca.gov/faces/codes.xhtml.

California Business & Professions Code [Cal BPC]. Retrieved July 26, 2014 from http://leginfo.legislature.ca.gov/faces/codes.xhtml.

California Code of Civil Procedure [Cal CCP] Part 2 Title 2 Chapter 3 §337.1. (1967). Retrieved August 11, 2014 from http://leginfo.legislature.ca.gov/faces/codes.xhtml

California Code of Civil Procedure [Cal CCP] Part 2 Title 2 Chapter 3 §337.15. (1981). Retrieved August 11, 2014 from http://leginfo.legislature.ca.gov/faces/codes.xhtml.

California Environmental Protection Agency Air Resources Board. *Laws and Regulations.* Retrieved August 15, 2014 from http://www.arb.ca.gov/html/lawsregs.htm.

California Labor Code [Cal LAB] Division 2 Part 4 Chapter 1 §1182.12. Retrieved August 16, 2014 from http://leginfo.legislature.ca.gov/faces/codes.xhtml.

California Occupational Safety and Health Regulations [Cal/OSHA], 8 California Code of Regulations Division 1 Chapter 3.2 Subchapter 2 Article 2.5 §§341.6 *et seq.* Retrieved August 15, 2014 from http://www.dir.ca.gov/dlse/ccr.htm.

Carbine v. Sutherlin, 544 So.2d 455 (4th Cir. 1989).

Civil Rights Act of 1964 [Civil Rights Act], 42 USC §§1971 *et seq.* (1964). Retrieved July 27, 2014 from http://uscode.house.gov.

Clean Water Act of 1977 [Clean Water Act], 33 USC §1251 *et seq.* (1977). Retrieved August 19, 2014 from http://uscode.house.gov.

Coe v. Thermasol Ltd. 615 F. Supp. 316 (W.D.N.C. 1985) aff'd 785 F.2d 511 (4th Cir. 1986).

Commonwealth v. John Hunt & others. 45 Mass. 111; 1842 Mass. Lexis 111; 4 Met. 111. (1842).

Constitution of the United States of America. (adopted September 17, 1787) Preamble.

Contract Disputes Act [CDA]. 41 USC §§601 *et seq.* (1978). Retrieved July 29, 2014 from http://uscode.house.gov.

Contract Work Hours and Safety Standards Act. 76 Stat. §357 (1962).

Copeland Anti-Kickback Act. 18 USC §847: Kickbacks from public works employees. (1948). Retrieved July 29, 2014 from http://uscode.house.gov.

Corporation of Presiding Bishop of Church of Jesus Christ of Latter-Day Saints v. Cavanaugh (1963) 217 Cal. App. 2d 492, 508 [32 Cal. Rptr. 144]

County of Cook v. Henry Harms. 108 Ill. 151 (1883).

Crawford Painting & Drywall Co. v. J.W. Bateson Co. Inc. 857 F.2d 981 (5th Cir. 1988).

Davis-Bacon Act. ch. 411, 46 Stat. §1494 (1931).

Department of Transp. V. IA Constr. Corp. 588 A.2d 1327 (Pa. Commw. Ct. 1991).

Downey v. Clauder 811 F. Supp. 338 (S.D. Ohio 1992).

Eichberger v. Folliard 169 Ill.App.3d 145, 523 N.E.2d 389, appeal denied, 122 Ill.2d 573, 530 N.E.2d 243 (1988).

Enlow & Son, Inc., v. Higgerson Brothers. 113 S.E.2d 855 (1960).

Fair Labor Standards Act of 1938 [FLSA], 29 USC §§201 *et seq.* (1938). Retrieved August 16, 2014 from http://uscode.house.gov.

False Claims Act. 31 USC §3729. (2014). Retrieved July 29, 2014 from http://uscode.house.gov.

False Statements Act, 18 USC 1001 (2014). Retrieved July 29, 2014 from http://uscode.house.gov.

Faragher v. City of Boca Raton, 524 US 775 (1998).

Farmers Export Co. v. M/V Georgis Prois, Etc. 799 F.2d 159 (5th Cir. 1986)

Federal Acquisition Regulations System [FAR], 48 CFR. (1984). Retrieved July 29, 2014 from Electronic Code of Federal Regulations [e-CFR] at http://www.ecfr.gov/cgi-bin/ECFR?page=browse.

Federal Arbitration Act. 9 USC §§1-14. (1925). Retrieved July 27, 2014 from http://uscode.house.gov.

George A. Fuller Co. v. United States 69 F.Supp.409 (Ct. Cl. 1947).

Georgia Port Authority v. Norair Engineering 195 S.E.2d 199 (Ga. Ct. App. 1973).

Hubbard Business Plaza v. Lincoln Liberty Life Ins. Co. 649 F. Supp. 1310 (D. Nev. 1986)

Jacob & Youngs, Inc. v. Kent, Court of Appeals of New York, 230 N.Y. 239; 129 N.E.889; 1921 N.Y. LEXIS 828; 23 A.L.R. 1429, December 1, 1920, Argued, January 25, 1921, Decided.

James Baird Co. v. Gimbel Brothers, Inc. 64 F.2d 344. (1933).

Labor-Management Reporting and Disclosure Act of 1959, 29 USC §§401 *et seq.* (1959). Retrieved August 16, 2014 from http://uscode.house.gov.

Labor Management Relations Act, 29 USC §§141 *et seq.* (1947). Retrieved August 16, 2014 from http://uscode.house.gov.

Lakeshore Engineering Services, Inc. v. the United States. 110 Fed.Cl. 230. (April 3, 2013). citing Liebherr Crane Corp. v. United States, 810 F.2d 1153 (Fed. Cir. 1987), Bromley Contracting Co. v. United States, 794 F.2d 669 (Fed. Cir. 1986), and United States v. Hamilton Enterprises, Inc. 711 F.2d 1038 (Fed. Cir. 1983).

Larry C. Lempke v. Brian Dagenais, 130 N.H. 782; 547 A.2d 290; 1988 N.H. Lexis 59.

Lee v. Red Lobster Inns of America, 92 Fed. Appx. 158 at 161 (6th Cir. 2004).

Leonard v. Home Builders (1916) 174 Cal. 65, 69 [161 P. 151].

Lewis v. Jones 251 S.W.2d 942 (Tex. Ct. App. 1952).

Lingenfelder v. Wainwright Brewery Co., 15 S.W. 844 (Mo. 1891).

Mannix v. Tryon (1907) 152 Cal. 31, 40 [91 P.983, 987]

Maxum Foundations, Inc. v. Salus Corp., 507 N.E.2d 588 (Ind.Ct.App.1987)

McCreary v. Shields, 333 Mich. 290. (1952).

McQuagge v. United States, 197 F. Supp. 460 (W.D. La. 1961).

Meco v. Dancing Bear, 42 S.W.3d 794 (Mo. 2001).

National Emission Standards for Hazardous Air Pollutants [NESHAP], 40 CFR §§61.140 *et seq.* (1984). Retrieved August 15, 2014 from Electronic Code of Federal Regulations [e-CFR] at http://www.ecfr.gov/cgi-bin/ECFR?page=browse.

National Labor Relations Act [NLRA], 29 USC §§151 *et seq.* (1935). Retrieved July 28, 2014 from http://uscode.house.gov.

National Railroad Passenger Corporation v. Morgan, 536 U.S. 101 (2002).

Native Homes, Inc. v. Stamm, 721 So.2d 809 (Fla. Dist. Ct. App. 1998).

Negligence. §1. *57 Am. Jur. 2d* (1971).

Norris-LaGuardia Act, 29 USC §§101 *et seq.* (1932). Retrieved August 16, 2014 from http://uscode.house.gov.

Office of Federal Compliance Program [OFCCP], 41 CFR Volume 1 Subtitle B Chapter 60. (1978). Retrieved August 20, 2014 from Electronic Code of Federal Regulations [e-CFR] at http://www.ecfr.gov/cgi-bin/ECFR?page=browse.

Occupational Safety and Health Act of 1970 [OSH Act], 29 USC §§651 *et seq.* (1970). Retrieved July 28, 2014 from http://uscode.house.gov.

Paradine v. Jane, Court of King's Bench (1642).

Polychlorinated Biphenyls [PCBs] Manufacturing, Processing, Distribution in Commerce, and Use Prohibitions, 40 CFR Part 761. Retrieved August 15, 2014 from Electronic Code of Federal Regulations [e-CFR] at http://www.ecfr.gov/cgi-bin/ECFR?page=browse.

Pilot program for enhancement of contractor protection from reprisal for disclosure of certain information, 41 USC §4712. Retrieved July 29, 2014 from http://uscode.house.gov.

Plaza Dev. Serv. V. Joe Harden Builders, Inc. 365 S.E.2d 231 (S.C. Ct. App. 1988).

Pollard v. Saxe & Yolles Dev. Co. (1974) 12 Cal.3d 374, 378-379 [115 Cal. Rptr. 648].

Rehabilitation Act of 1973. 29 USC §§701 *et seq.* (1973). Retrieved August 20, 2014 from http://uscode.house.gov.

R.L. Jones *et ux.* v. J.H. Hiser Construction Co., Inc. 60 Md. App. 671, 484 A.2d 302. (Md. Spec. App. 1984).

Santucci Construction Co. v. Cook County. 21 Ill.App.3d 527. (1974).

Sarbanes-Oxley Act, P.L. 107-204, 107th Cong., 116 Stat. 745 (2002) (enacted).

Schnell v. Nell, 17 Ind. 29 (1861).

Scott v. Strickland 691 P.2d 45 (Kan. Ct. App. 1984).

Sea Box, Inc. B-291056, 2002 U.S. Com. Gen. Proc. Dec. P181.

Senate Bill 800. California Civil Code §§ 896 et seq. (2003).

Shoals v. Home Depot, Inc., 422 F.Supp.2d 1183 (2006).

Siders v. Schloo (1987) 188 Cal. App.3d 1217 [233 Cal. Rptr. 906].

Small Business Act, 15 USC §§631 *et seq.* (1958). Retrieved August 20, 2014 from http://uscode.house.gov.

Small Business Reauthorization Act of 1997. 15 USC §§631 *et seq.* (1997). Retrieved August 20, 2014 from http://uscode.house.gov.

Snyder Plumbing & Heating Corp. v. State 198 N.Y.S.2d 600, 604 (Ct. Cl. 1960).

Somerville v. Jacobs, 153 W.Va. 613. (1969).

Southeastern Land Fund, Inc. v. Real Estate World, Inc. 227 S.E.2d 340 (Ga. 1976).

Sulzer Bingham Pumps, Inc. v. Lockheed Missiles & Space Co., Inc. 947 F.2d 1362 (1991).

Sunbeam Construction Co. v. Fisci (1969) 2 Cal. App.3d 181, 184 [82 Cal. Rptr. 446 Cal. App. 1969].

T. Brown Constructors, Inc. v. Pena 132 F.3d 724 (Fed. Cir. 1997) .

The Miller Act of 1935, 40 USC §§3131 *et seq.* (1935). Retrieved August 11, 2014 from http://uscode.house.gov.

The Sherman Antitrust Act of 1890 [Sherman Act]. (1890). 15 USC §§1-7. Retrieved July 29, 2014 from http://uscode.house.gov

Thomas v. Thomas' Executor, 55 Ky. (16 B. Mon.) 420.

Toxic Substances Control Act [TSCA]. 15 USC §§2601 et. seq. Retrieved August 15, 2014 from http://uscode.house.gov.

Trade Agreements Act of 1979. 19 USC §§2501 *et seq.* (1979). Retrieved August 20, 2014 from http://uscode.house.gov.

U.S. v. Spearin, 248 U.S. 132 (1918).

U.S. v. Young Lumber, 376 F.Supp. 1290 (D.S.C.1974).

Unites States f/b/o Wallace v. Flintco, Inc. 143 F.3d 955 (5th Cir. 1998).

Vietnam Era Veterans Readjustment Assistance Act of 1974. 38 USC §§101 *et seq.* (1974). Retrieved August 20, 2014 from http://uscode.house.gov.

Voss v. Forgue, Fla., 84 So. 2d 563. (1956).

W.G. Cornell Co. v. United States 376 F.2d 299 (Ct. cl. 1967).

Walters v. Metropolitan Educational Enterprises, Inc., 117 S. Ct. 660 (1997).

Western States Constr. Co. Inc. v. United States 26 Cl.Ct. 818 (1992).

William A. Drennan v. Star Paving Company, 51 Cal.2d 409. (1958).

INDEX

A

acceleration, 79
acceptance, 9, 10, 37, 58, 109, 110, 183, 184
ACM, 233–37
addenda, **27**, 86, 135, 137, 212
administration of the contract, 37–38
administrative law judge [ALJ], 5, 255
ADR, 115, 118, 207
adverse impact, 273, **276**
affidavit, **52**, 170
affirmative action, 162, 219, 228, **231**, 270, 271, 274
AFL, 248
AFL-CIO, 249
AGC, 3, 6, 23
agency shop, 266
agreement, 25
agreements to agree, 10, **17–19**
AIA, 3, 23, 25, 26, 64, 65, 86, 115, 117, 119, 125, 132, 289
all risk insurance, 202
ally doctrine, 264
alternates, 88–89, **134–35**, 153
alternative dispute resolution. *See* ADR
ambiguity, 87
American Arbitration Association, 121, 123
American Federation of Labor. *See* AFL
American Institute of Architects. *See* AIA
American Subcontractors Association, Inc. [ASA], 6
Anti-Kickback Act of 1986, 175
Antitrust Laws. *See* Sherman Antitrust Act
apparent mistake, 148–49
application for payment, 37, **42–44**, 51, 92–93, 98, 101, 164, 167
arbitration, 74, 115, **118–23**, 125–26, 161, 288
arbitration agreement, 288–89
architect. *See* designer
Armed Services Board of Contract Appeals [ASBCA], 127

ASA, 185–86
asbestos-containing material. *See* ACM
as-built drawings, 33, 51
assignment, 32, **39–40**, 101–2, 157–58
Associated Builders and Contractors, Inc. [ABC], 6
Associated General Contractors of America. *See* AGC
Associated Specialty Contractors, Inc. [ASC], 6
assumption of risk, 194
authorization cards, 257
automobile insurance, 203

B

bargaining unit, 252, **256**, 258–59
best value procurement, 138–39
bid bond, 137, **211–12**, 215
bid depository, 182
bid guarantee, 137, **211–12**
bid peddling, 174, **181–86**
bid protest, 133, **153–55**
bid rigging, 176
bid rotation, 176
bid security, 137
bid shopping, 174, **181–86**
bid suppression, 176
bilateral agreements, 11
Board of Contract Appeals, 127–28, 149, 153–54
boiler and machinery insurance, 202
bona fide occupational qualification [BFOQ], 274
bond of qualifying individual, 224
bond premium, 216
bridging, 141
broad form indemnification. *See* indemnification
builders risk insurance, 187, 199, **201–3**
burial grounds, archaeological sites, and wetlands, 62
business agent, 267
Buy America Act of 1988, 232

C

Cal-EPA, 236
Cal-EPA Air Resources Board, 236
California Department of Consumer Affairs, 221
California Environmental Protection Agency. *See* Cal-EPA
California Indoor Radon Program, 237
California lead-based paint rules, 241
Cal-OSHA, 235
captive-audience speeches, 258
cardinal change, 16, 64–66
cases, 6
certificate for payment, **42–44**, 52, **92–94**, 98, 99
certificates of insurance, 52, 200, 212
CGL, 35, 187, **199–201**, 202, 204
change order, 27, 33, 36–39, 44, 47, **65–68**, 65–68, 74–79, 84, 86, 110–11, 151, 162, 182, 201–2, 209
changed conditions. *See* concealed or unknown conditions
changes clause, **63–65**, 166
Cigarmakers' Union. *See* AFL
CIO, 249
citation, 6
Civilian Board of Contract Appeals [CBCA], 127
claim, 31, 36, 43, 49, 52, 57, 60, 62, 64, 65, 74–79, 83, 87–89, 91, 94, 104, 112–13, **115–28**, 132–34, 153–55, 159–60, 166–71, 173–75, 177, 182, 196, 197–200, 204–5, 211–13, 226, 269, 272, 274, 276, 278, 279, 286–89
claim document, 115–16
claims-based policy. *See* CGL
class-action lawsuit, 287
Clean Water Act, 234, 245
clerical mistake, 148–50
closed shop, 266
CM, 124, **142–45**, 145, 209, 221, 270
CM-agency [CMa], 143–45
CM-at-risk [CMAR], 143–45
CO, 58, 127, 131, **133–36**, 147–52, 152, 154, 212, 229, 232
code of ethics. *See* ethics
collusion, 138, 174, **176**

commercial general liability insurance. *See* CGL
common law, **4–9**, 13, 41, 56, 60, 63–64, 66, 67–68, 72, 75, 85, 104, 108, 183, 196–97, 233, 269
common situs picketing, 264
comparative fault, 193
comparative form indemnification. *See* indemnification
comparative liability, 196
compensable delay, **76–79**, 87, 89, 103, 211
compensatory time off, 280
competitive sealed bidding, **137–38**, 147–48
complementary bidding, 176
concealed or unknown conditions, **55–62**, 76, 91, 131, 160
concurrent delay, 76–77
conditions, 25
conflict of interest, 174, **178–79**
Congress of Industrial Organizations. *See* CIO
ConsensusDocs™, 3, 23, 25, 26, 86, 118
consent election, 257
consequential damages, **104**, 192
consideration, **6–12**, 63–64, 67–68, 151, 183
consolidation, 122, 125–26
construction change directive, 37, **65–67**, 74, 76, 79, 110
Construction Industry Arbitration Rules, 123
construction management. *See* CM
Construction Specifications Institute. *See* CSI
constructive change, 110–11
constructor. *See* contractor
contingent assignment, 32
continuous delays, 74
contra proferentum, 87
Contract Disputes Act, 127, 132
contract documents, 25–28
contract of adhesion, **163**, 170, 288
Contract Work Hours and Safety Standards Act, 228
contract-bar rule, 259
contracting officer. *See* CO
contractor, 29, **31–36**, 224

contractor licensing, 219–28
Contractor's State License Board. *See* CSLB
contractor-subcontractor agreements, 65, 227
contributory negligence, 193
Copeland Anti-Kickback Act, 175
cost-plus, 45–48
cost-plus-a-percentage-of-cost, 47
cost-plus-fixed-fee, **45**, 47, 145
cost-plus-fixed-percentage, 46
cost-plus-variable-percentage, 46
cost-reimbursement incentive contracts, 48
counter-offers, 9
course of dealing, 82–83
course of performance, 82–83
Court of Federal Claims, 127, 148–49, 153–55
criteria documents, **141–42**, 144
critical activity. *See* critical path
critical path, **74**, 76–77, 103, 124
CSI, 88–89
CSI 3-part section format, 89
CSLB, 5, 221–24

D

damages, 13–15
damages-type bond, 212
date of commencement, 25, **74–75**
Davis-Bacon Act, 162, **228**
DB, **141–42**, 144–46, 146
DBB, **140**, 141–43, 145, 147
debarment, **136**, 229
deductive change, 44, 63, **66–67**, 113, 151, 162, 166
defect, 19, 31, 38, 39, 46, 52, 57, 88, **105–12**, 107, 188, 212–13, 214, 219–21, 225
demonstrative evidence, 116
deposition, 123–24
design specifications. *See* prescriptive specifications
design-bid-build. *See* DBB
design-build. *See* DB
designer, 29, **37–38**
differing conditions. *See* concealed or unknown conditions

differing site conditions. *See* concealed or unknown conditions
direct evidence, 273–74
disadvantaged business enterprises [DBEs], 230
disclaimers, 61
discovery, 123–26
discrimination, 287
disparate treatment, 273–76
dispute resolution, **115–26**, 131, 160–61, 198, 202, 207
dispute review board, 118–20
documentary evidence, 116, 213
donative promise, **7**, 8, 183
drawings, 25
duty of cooperation, 21

E

e-bidding, **147–48**, 185
economic losses, 191
economic price adjustment, **45**, 140, 141
economic strike, 261
economic waste, 14
Eichleay formula, 78–79
EJCDC, 3, 23, 25, 26, 64, 118, 132, 290
employment contract, 179, 180, 269, **270–71**, 288
employment discrimination, 179, 181, 231, 253, 269, 270, **272–79**
employment practices, 174, 179–80, 274–75
employment-at-will, **270–72**, 278
encumbrance, 168
engineer. *See* designer
Engineers Joint Contract Documents Committee. *See* EJCDC
entitlement, 228–32
environmental law, 233–45
EPA, 5, 235–45
Equal Pay Act, 269, **278–79**
equipment floater insurance, 35, 187, 199, **203**
equitable, 58, **60**, 76
equivalent lump sum, 49
errors or omissions, 21, 22, 33, 76, 89, 103, 147, 199
ethics, 135, **173–86**

Excelsior list, 257
exculpatory clause, 79
excusable delay, **74–75**, 76–77, 79, 103, 110, 198
Executive Orders 10988 and 11491, 252
exempt, 279
expectation damages, **11**, 14
expiration, 9, 200, 202, 259
express warranty, 19, 105, **106**

F

face value, 208
facially discriminatory, 273–74
failure to reasonably accommodate, 273, **277**
Fair Labor Standards Act. *See* FLSA
False Claims Act, 175
False Statements Act, 174–75
false statements in bids, 174
FAR, 44–45, 47–48, 58–60, 128, **132–55**, 173–75, 178–79, 181, 184, 212, 228–29, 231
fast tracking, 142
Federal Acquisition Regulation. *See* FAR
Federal Arbitration Act, 122, 161, **288**
fiduciary, 46–47, 116, 143–44, **197–98**, 203
final certificate for payment, 52
final completion, 38, 42, 43, **51–53**, 105, 108, 142, 200, 170, 200, 211, 225
final lien waiver. *See* lien waiver
firm-fixed-price. *See* fixed-price
fixed-price, **44–48**, 145
fixed-price incentive contracts, 48
flow-thru clause, 39, **160–61**
FLSA, 269, **279–82**
force majeure, 74, **75–76**, 77
forfeiture-type bond, 212
forty-eight hour rule, 257
friable asbestos, 235
front-loaded bids, 42, **135**
fully integrated agreement, 24

G

general aggregate. *See* limits of liability
general conditions, **26**, 92, 105, 132
general contractor. *See* contractor
general requirements, 27–28, **78**
German Workingmen's Union, 248
GMP, 46, **143–45**
good faith and fair dealing, **20–21**, 22, 46, 71, 87, 89, 131
Government Accountability Office [GAO], 153–54
Government Services Administration, 127
guaranteed maximum price. *See* GMP
guarantor, 207–8

H

harassment, 272, 273, **276**
hiring hall, 267
Historically Underutilized Business Zone. *See* HUBZone
hot cargo clause, 265
HUBZone, 231

I

IFB, **27**, 137, 138, 139
immaterial breach. *See* minor breach
implied warranty, 19–20, 57, 90, 105
implied-in-fact contracts, 11
incentives, 46, 48, 141, 258
inchoate contract, **18**, 71
incorporation by reference, 39, **160–61**
indemnification, **195–97**, 199, 200, 216
indemnification agreement, 216
indemnify. *See* indemnification
indemnitee. *See* indemnification
indemnitor. *See* indemnification
indemnity. *See* indemnification
Industrial Workers of the World. *See* IWW
initial decision maker, 117–18
injunction, **14**, 250, 265, 266
instruments of service, 51–52

insurance, 25, 28, 35, 52, 91, 102, 117, 131, 160, **187–205**, 208, 209, 211–12, 224, 228, 235
integrated-design-bid-build [IDDB], 144–45
interested party, 153–54
interim contract, 17, **18**
intermediate form indemnification. *See* indemnification
International Chamber of Commerce, 123
International Workingmen's Association, 248
invitation for bids. *See* IFB
irrevocable bid, 183
IWW, 248

J

job order contracting, 146–47
job steward, 267
joinder, 122–23, 125–26
jurisdictional picketing, 265
jury trial, 122, **124–25**

K

Knights of Labor, 248

L

labor agreement, 45, 259, 261, 265, **266–68**, 271
labor dispute, 75, **250–51**, 255, **261–68**, 287
labor dispute picketing, 262
labor law, 247–68
laboratory conditions, 258
Landrum-Griffin Act, 254
latent ambiguity, 87
latent defect, **108**, 111–12, 221
LBP, 55, 234, **238–43**
lead-based paint. *See* LBP
Lead-Based Paint Renovation, Repair and Painting Program. *See* RRP
legal compliance, 174, 180, 269
legal condition, **26**, 92
letter of intent, **17–18**, 23–24
liability, 56–57, 188–205, **188**

license bond, **212–13**, 224
lien claim. *See* mechanics' lien
lien release bond, 213
lien waiver, 170
limitation period, 287
limits of liability, 28, **200**, 204
line-item retention, 166
liquidated damages, 25, **53**, **72–73**, 79, 91
listing statutes, 184
litigation, 77, 79, 115–23, **123–26**, 178, 207, 235, 255
lockout, 261

M

market allocation, 176
mass picketing, 262
MasterFormat™, **27**, 28
material and workmanship specifications. *See* prescriptive specifications
material breach, **49–53**, 69, **94–97**, 103, 107, 207–9, 229, 270
material deviation, 133–35
mechanics' lien, 30, 32, 43, 70, 99, 100, **166–71**, 226
mediation, 74, 115, **117–22**
meeting of the minds, 10
memorandum of understanding [MOU], 17, **18**
memorializing, 12
merger clause, 24–25
merit shop, 266
method and material specifications. *See* prescriptive specifications
Miller Act of 1935, **208**, **210–11**
minimum wage, 272, **279–81**
mini-trial, 118, **120**
minor breach, **49**, **94–97**, 209
minor change, **65–67**, 76, 110
minority business enterprises [MBEs], 230
mistakes in judgment, 148–49
mixed motive, 273, **275**
modification, **27**, 45, 87, 93, 205, 209
modified comparative fault, 194
motion for entry upon land, 124
mutual assent, 10

307

mutual mistake, 150–52

N

named insureds, **197**, 199, 200
named perils insurance, 202
narrow form indemnification. *See* indemnification
national agreement, 267
National Emission Standards for Hazardous Air Pollutants. *See* NESHAP
National Industrial Recovery Act [NIRA], 251
National Institute of Occupational Safety and Health [NIOSH], 283
National Labor Board [NLB], 251
National Labor Relations Act. *See* NLRA
National Labor Relations Board. *See* NLRB
negligence, 79, **187**, 188–97, 197, 269
negotiated procurement, **139**, 147
NESHAP, 235–36
NLRA, **250–65**
NLRB, 5, 252, **254–56**, 257–66
NLRB Office of the General Counsel, 255
no-damage-for-delay, 79–80
no-lien contract. *See* mechanics' lien
noncompensable delays, **75**, 103
nonexempt, 279
nonresponsive, 185
nonresponsive bidder, **133–35**, 148, 153, 176, 212, 229
Norris-LaGuardia Act, 250, 261
no-strike, no-lockout provisions, 261
notary public, 52
notice of mechanics' lien. *See* mechanics' lien
notification to proceed, **74**, 155
novation, **39**, 154, 158

O

obligee, 207–8
Occupational Safety and Health Act of 1970. *See* OSH Act
Occupational Safety and Health Administration. *See* OSHA
Occupational Safety and Health Review Commission [OSHRC], 283, 285
occurrence-based policy. *See* CGL
offer, 9, 10, 183, 184, 214, 215, 216, 266, 271
one-year correction period, 112–13
open season, 259
open shop, 266
opportunity wage, 279
or approved equal, 89
or equal, 82, **89**, 134
order of precedence, 85–86
ordinances, 5, 33, 100
OSH Act, 179, 204, **283–86**
OSHA, 204, **283–89**
OSHA abatement period, 284
OSHA citation, 284
OSHA inspections, 284–85
OSHA standards, 284
output specifications. *See* performance specifications
owner, 29–31
owner-contractor agreement, **25**, 67, 74, 86, 91, 99–100, 209, 211–12, 227

P

parol evidence rule, **24**, 27, 82
partial lien waiver. *See* lien waiver
partially integrated agreement, **25**
participation and opposition, 277
patent ambiguity, 87
patent defect, **108**, 111–12
pattern or practice, 275
pay-if-paid clause, 165
payment bond, 91, 101, 171, 208, **210–12**, 214, 215
pay-when-paid clause, **165**, 177
penal sum. *See* face value
per diem, 53, **72–73**, 78
perfecting a lien. *See* mechanics' lien
performance, 32–33
performance bond, 66, 101, **208–9**, 212, 214
performance evaluation, contractor, 136
performance specifications, 35, **90**, 107, 140

picketing, 261, **262-63**, 263-65
polychlorinated biphenyls [PCBs], 243
post-award bid shopping, 182
pre-award bid shopping, 182
pre-bid conference, 133, 185
preconstruction services, 142-45
preexisting duty rule, 63, 67-68
pre-hire agreement. *See* union security clause
preliminary contract, 17, **18**
preliminary notice. *See* mechanics' lien
prequalification, **141**, 214-15
prescriptive specifications, 45, **88-89**, 107, 140
presumptive acceptance, 184
pretext, 273, **275**
prevailing wages, 228
price fixing, 176
primary contractor, 263-64
prime contractor. *See* contractor
prime trade contractor. *See* subcontractor
principal, 188, 207-8, 209-16
privity, 28
pro rata, 78, 168, **196**
procedural unconscionability. *See* unconscionability
product data, 34
professional stature, 180
progress payments, **41-42**, 70, 164-65, 170
project concession agreement, 145
project labor agreement, 267
project manual, **27**
project neutral, 118-19, **120**
promissory estoppel, 152, **183-84**
proprietary specification, 89
protected class, 231, **272-76**
protected interest, 191
public agencies, 31, 47, 109, **131-55**, 185-86, 271, 286
public construction contracts, 44, 131-39, 164, 185, 211
public private partnership [P3], 145-46

Q

qualifier. *See* qualifying individual

qualifying individual, 222-24
quantum meruit, **16**, 227
quid pro quo, 6-9, 11

R

radon, 237-38
radon-resistant building techniques, 238
recognitional picketing, 262, 265
regulations, 5
regulatory agencies, **5-6**, 62
rejection, 9, 110, 137
release of lien. *See* mechanics' lien
remedies, 13-16, 108, 160, 176, 181, 207, 251, 259, 265-66, 271
reporters, 6
representation election, 252, 254, **257-59**, 258, 262
request for production of documents and things, 124
request for proposals. *See* RFP
request for qualifications [RFQ], **141**, 146
responsible bidder, 133, **135-38**, 153, 181, 183, 184, 185, 212
responsible managing employee. *See* RME
responsible managing officer. *See* RMO
responsive bidder, **133**, 137, 150, 153, 181, 183, 184, 185, 212
restitution, 11, **14-17**, 151, 176, 226-27
results specifications. *See* performance specifications
retainage. *See* retention
retaliation, 272, 273, **277-78**
retention, 25, 42, 43, **165-66**, 170, 177
reverse bid auction, 185-86
reverse discrimination, 274
revocation, 9, 221
RFP, **27**, 139, 141, 145
right to sue letter, 287
right to work states, 266
risk premium, 111, 146, 193
RME, 223-24
RMO, 223-24
RRP, 239-43
RSMeans™, 146

S

safety practices, 35, 124, 174, 179
SBA, 229–30
schedule of values, 42, 92, 135
SDB, 230–31
SDVO, 230–31
security interests, **70–71**, 73, 167–68
service-disabled veteran-owned business enterprise. *See* SDVO
set-asides, 138, 146, 228–**31**
Sherman Antitrust Act, 176, **250**
shop drawings, 33–34
slowdown, 262
Small Business Administration. *See* SBA
Small Disadvantaged Business. *See* SDB
social contract, 179
Socialists, 248
sovereign immunity, 195
Spearin doctrine, **57**, **88–90**
specialty contractor. *See* subcontractor
specific performance, **14**, 35, 125
specifications, 25
standing, 153, 197
standing neutral, 120
Statewide Investigative Fraud Team [SWIFT], 224
statute of frauds, 13
statute of limitations, 115, 154, 199, 200
statutes, 5
statutory codes, 5–6
statutory law, **5**, 6, 70, 266, 269
stop notice, 169
storm water pollution prevention plan [SWPPP], 245
strict liability, **192–93**, 204
strike, 83, 247–48, 250, 251–52, **261–62**, 262–64, 266
subcontractor, 29, **38–39**
submittals, **33–35**, 37
subrogation, **197**, 203
substantial completion, 25, 38, 42, 48, **50–51**, 53, 74, 77, 108, 170, 199
substantial compliance, 227–28
substantial performance, 49–51
substantive unconscionability. *See* unconscionability

substitution, **65**, 83, **89**, **134**, 159
summary jury trial, 118, **120–21**
supplementary conditions, 132
supporting documents, 51–52
surety bond, 52, 131, 137, 160, 198, 202, **207–16**
surety investigation, 214–15
suspension, 76, **97–103**, 116, 131, 136, 160, 221–22, 229, 278

T

Taft-Hartley Act, 253
target cost, **46**, **48**, 145
target housing, 239–42
technical specification, **27**
temporary contract, 17, **18**
termination, 19, 53, **95–103**, 107, 116, 131, 151, 153, 160, 229, 245, 270
The "Old" National Labor Relations Board, 252
time is of the essence, 69–73
Toxic Substances Control Act. *See* TSCA
Trade Agreements Act of 1979, 232
trade contractor. *See* subcontractor
trade financing, 158
TSCA, 233, 240
twenty-four hour silent period, 258
type I differing conditions, 59
type II differing conditions, 60

U

U.S. Justice Department, 180
UCC, **68**, 109–10
umbrella excess liability insurance, 187, 199, 204
unconscionability, 148, 288–89
unencumbered equity. *See* encumbrance
unfair labor practice, 253–55, 259–62, **259**, 261–62, 265
unfair labor practice strike, 261
Uniform Commercial Code. *See* UCC
unilateral agreements, 12
unilateral mistake, **148–49**, 150
union local, 267
union security clause, 266
union shop, 266

United States Environmental Protection Agency. *See* EPA
unit-price, 44, **48–49**, 135, 140
unjust enrichment, 11, **15**, 226
usage, 82–83
usage of trade, 18, 82–83

V

verbal contracts, 10, **12–13**
veteran-owned business enterprises [VBEs], 230
vicarious liability, 38, **193–95**
violent picketing, 261, 262
voluntary recognition, 257

W

Wagner Act. *See* National Labor Relations Act [NLRA]
waiver of subrogation, 203
warranty, 19–20, 28, 32–33, 52, 57, 90, **105–13**, 115, 131, 157, 163, 187–88, 208, 211
warranty of fitness, **20**, 131
warranty work, **111–13**, 188, 211
whistleblower, 181
wildcat strike, 261
Wobblies. *See* IWW
women-owned business enterprises [WBEs], 230
work product exclusion, 188
workers' compensation insurance, 187, 199, **204–5**, 204, 224
workmanlike performance, 19, 107, 131

Y

yellow-dog contracts, 248, **250–51**